Millimeter Wave Radar

Christian Waldschmidt · Christina Bonfert ·
Timo Grebner

Millimeter Wave Radar

Hardware and Signal Processing

Christian Waldschmidt ⓘ
Universität Ulm
Ulm, Baden-Württemberg, Germany

Christina Bonfert
Universität Ulm
Ulm, Baden-Württemberg, Germany

Timo Grebner
Universität Ulm
Ulm, Baden-Württemberg, Germany

ISBN 978-3-031-89117-5   ISBN 978-3-031-89118-2 (eBook)
https://doi.org/10.1007/978-3-031-89118-2

This work was supported by Universität Ulm.

© The Editor(s) (if applicable) and The Author(s) 2025. This book is an open access publication.

Open Access This book is licensed under the terms of the Creative Commons Attribution 4.0 International License (http://creativecommons.org/licenses/by/4.0/), which permits use, sharing, adaptation, distribution and reproduction in any medium or format, as long as you give appropriate credit to the original author(s) and the source, provide a link to the Creative Commons license and indicate if changes were made.
The images or other third party material in this book are included in the book's Creative Commons license, unless indicated otherwise in a credit line to the material. If material is not included in the book's Creative Commons license and your intended use is not permitted by statutory regulation or exceeds the permitted use, you will need to obtain permission directly from the copyright holder.
The use of general descriptive names, registered names, trademarks, service marks, etc. in this publication does not imply, even in the absence of a specific statement, that such names are exempt from the relevant protective laws and regulations and therefore free for general use.
The publisher, the authors and the editors are safe to assume that the advice and information in this book are believed to be true and accurate at the date of publication. Neither the publisher nor the authors or the editors give a warranty, expressed or implied, with respect to the material contained herein or for any errors or omissions that may have been made. The publisher remains neutral with regard to jurisdictional claims in published maps and institutional affiliations.

This Springer imprint is published by the registered company Springer Nature Switzerland AG
The registered company address is: Gewerbestrasse 11, 6330 Cham, Switzerland

If disposing of this product, please recycle the paper.

Preface

Radar sensor technology has undergone an impressive transformation over the last 20 years. Originally specialized for long-range applications in the microwave range, radar sensors in the millimeter wave range have developed into a versatile and technologically mature sensor technology. The modern use of radar sensors in the millimeter wave range differs markedly from the traditional approaches discussed in the classic radar literature. This is exactly where our book comes in: We want to close a gap in the existing technical literature and comprehensively present modern radar sensor technology in the millimeter wave range.

In this work, we discuss current technologies and methods such as chirp sequence and digital modulation in detail and show how digital beamforming and MIMO systems can be implemented. Topics such as pulse modulations, RCS signatures or detection probabilities play only a minor role for the new applications; here we refer to the classic radar literature.

Our approach is holistic: we look at radar sensors from a system perspective and explain the interactions between hardware design and signal processing. We are convinced that high-performance radar sensors can only be realized by co-designing these two domains. Therefore, we start with the basics and functions of radar sensors in Part I of the book. This is followed by detailed considerations of the design of the signal processing and hardware in Parts II and III, respectively. For those who want to delve deeper into the subject matter, Part IV of the book offers an introduction to modern methods of environment perception and modeling as well as other practical topics.

The contents of the book are based on our experience in the industrial development of millimeter wave radar sensors as well as university research from the basics to industry-related applied research at the University of Ulm, Germany.

We would like to thank everyone who supported us in writing this book. Many thanks go to the companies that provided us with sample materials and to the proofreaders for their accurate checks. We would especially like to thank the many Ph.D. students and postdocs who have supported and assisted us with their research work.

We hope that this book will expand your knowledge and inspire you to make creative use of the many possible applications of modern radar sensor technology. We hope you enjoy reading this book and gain many new insights!

Ulm, Germany
2024

Christian Waldschmidt
Christina Bonfert
Timo Grebner

Competing Interests The authors have no competing interests to declare that are relevant to the content of this manuscript.

Contents

Part I Fundamentals

1 Introduction 3
   1.1 Frequency Bands and Areas of Application 4
   1.2 Radar Equation 6
       1.2.1 Radar Equation for Point Targets 6
       1.2.2 Radar Equation for Infinitely Extended Area Targets 9
       1.2.3 Monostatic and Bistatic Radars 10
   1.3 Radar Cross-Section (RCS) 12
   1.4 Measurands 14
       1.4.1 Range 14
       1.4.2 Relative Radial Velocity 15
       1.4.3 Micro Doppler 19
       1.4.4 Spatial Direction 19
       1.4.5 Other Measurands 20
   1.5 Wave Propagation 20
       1.5.1 Wave Propagation in the Atmosphere 21
       1.5.2 Multipath Propagation 21
       1.5.3 Clutter 23
   References 24

2 Applications 25
   2.1 Automotive Radar 25
   2.2 Radars for Rail, Air and Water Vehicles 27
   2.3 Sensors for Mobile Robotics 28
   2.4 Radars in Automation Technology 28
   2.5 Civil Security Technology 29
   2.6 Quality Assurance 30
   2.7 Medical Applications 31
   References 32

3 Resolution, Separability, Accuracy and Unambiguity of a Radar 33
- 3.1 Preliminary Mathematical Considerations 33
- 3.2 Introduction to Resolution, Separability, Accuracy and Unambiguity 35
- 3.3 Resolution 37
  - 3.3.1 Range Resolution 38
  - 3.3.2 Velocity Resolution 40
  - 3.3.3 Angle Resolution 41
- 3.4 Separability 43
- 3.5 Accuracy 44
- 3.6 Unambiguity 46
- References 47

Part II Radar Types, Modulation Schemes and Radar Imaging

4 Radar Types and Modulation Schemes 51
- 4.1 Analog Modulation Schemes 52
  - 4.1.1 CW Radars 52
  - 4.1.2 FMCW Radars 56
  - 4.1.3 Chirp Sequence Radars 64
- 4.2 Digital Modulation Schemes 69
  - 4.2.1 OFDM Radars 70
  - 4.2.2 PMCW Radars 76
- References 81

5 Signal Processing Principles 83
- 5.1 Key Quantities and Terms 85
- 5.2 Windowing 88
- 5.3 Range and Velocity Estimation 91
  - 5.3.1 Evaluation of CW Radars 92
  - 5.3.2 Evaluation of FMCW Radars 93
  - 5.3.3 Evaluation of Chirp Sequence Radars 95
  - 5.3.4 Evaluation of OFDM Radars 99
  - 5.3.5 Evaluation of PMCW Radars 105
- 5.4 Target Detection using CFAR Algorithms 109
  - 5.4.1 Cell Averaging CFAR (CA-CFAR) 111
  - 5.4.2 Ordered Statistics CFAR (OS-CFAR) 113
  - 5.4.3 CFAR Target Detection 114
- References 115

6	**Fundamentals of Antennas and Antenna Arrays**		117
	6.1	Antenna Array Parameters and Coordinate Systems	118
	6.2	Fundamental Considerations for Antenna Arrays	119
	6.3	Description of Antenna Arrays Using a Signal Model	123
	6.4	Calibration and Hardware Influences	125
	6.5	Beamformer and Power Evaluation	125
	6.6	Angular Resolution and Angular Separability	127
		6.6.1 One-Dimensional Antenna Arrays	127
		6.6.2 Two-Dimensional Antenna Arrays	129
	References		131
7	**Methods for Angle Estimation**		133
	7.1	Angle Estimation by Correlation	133
	7.2	Angle Estimation by Fourier Transform	134
	7.3	Angle Estimation Algorithms	135
		7.3.1 Bartlett Beamformer	136
		7.3.2 Capon Beamformer	137
		7.3.3 High-Resolution Methods: MUSIC	137
	7.4	Comparison of the Methods	141
8	**MIMO Radars and Antenna Array Design**		143
	8.1	Virtual Aperture	144
	8.2	Generation of Orthogonal Signals	149
		8.2.1 Time Division Multiplexing	149
		8.2.2 Frequency Division Multiplexing	151
		8.2.3 Code Division Multiplexing	152
	8.3	Phased Arrays and Performance Analysis for MIMO Radars	153
	8.4	Array Design and Evaluation	155
	8.5	Compensation of Near-Field Effects	157
	References		160

Part III Radar Hardware

9	**Hardware and Technology**		163
	9.1	System Partitioning	163
	9.2	Assembly and Interconnect Technology for Antennas and MMICs	164
		9.2.1 Material and Geometry Requirements	165
		9.2.2 Millimeter Wave Printed Circuit Boards	166
		9.2.3 Millimeter Wave Packages	167
		9.2.4 Transmission Lines and Transitions	168
	9.3	Antennas	170
		9.3.1 Patch Antennas	170
		9.3.2 Waveguide Antennas	171
		9.3.3 Integrated Antennas and Lenses	172

	9.4	MMIC and Technology Selection	174
	9.5	Signal Synthesis in Millimeter Wave Systems	175
	Reference		176
10	**Hardware Effects on System Level**		**177**
	10.1	Link Budget	177
	10.2	Leakage and Short-Range Behavior	178
	10.3	Phase Noise	180

Part IV Advanced Radar Topics

11	**Radar-Based Grid Maps**			**185**
	11.1	Amplitude Grid Map (AGM)		187
	11.2	Probabilistic Occupancy Grid Map (OGM)		190
		11.2.1	Local Map	190
		11.2.2	Map Update	193
	11.3	Comparison of AGM and OGM		193
	11.4	Simultaneous Localization and Mapping (SLAM)		196
		11.4.1	Dead Reckoning Based on Ego-Motion Estimation	196
		11.4.2	Scan Matching	199
	References			201
12	**Synthetic Aperture Radar (SAR) for Millimeter Wave Applications**			**203**
	12.1	Fundamentals and Resolution		204
	12.2	SAR Processing		206
	12.3	Probabilistic SAR Processing		208
	12.4	SLAM for SAR Applications		212
	References			212
13	**Coexistence and Interference of Radar Sensors**			**213**
	13.1	Interfering Signals		213
	13.2	Impact of Interference		214
		13.2.1	Impact on Analog FMCW and Chirp Sequence Radars	214
		13.2.2	Impact on Digital Radars	216
	13.3	Countermeasures		217
		13.3.1	Measures in the Time Domain	218
		13.3.2	Measures in the Time-Frequency Domain	219
		13.3.3	Measures in the Frequency Domain	219
		13.3.4	Measures in the Angular Domain	221
	References			222
Index				**223**

Abbreviations

ADC	Analog to digital converter
AGM	Amplitude grid map
AiP	Antenna in package
AIT	Assembly and interconnect technology
AoC	Antenna on chip
BP	Backprojection
CA-CFAR	Cell averaging CFAR
CDM	Code-division multiplexing
CFAR	Constant false alarm rate
CMOS	Complementary metal-oxide-semiconductor
CRLB	Cramèr-Rao lower bound
CUT	Cell under test
CW	Continuous wave
DAC	Digital to analog converter
DC	Direct current
DDS	Direct digital synthesis
DFT	Discrete Fourier transform
DoA	Direction of arrival
DTFT	Discrete-time Fourier transform
EIRP	Equivalent isotropic radiated power
eWLB	Embedded wafer level ball grid array
FDM	Frequency-division multiplexing
FFT	Fast Fourier transform
FMCW	Frequency-modulated continuous-wave
FoV	Field of view
FPGA	Field-programmable gate array
FSK	Frequency-shift keying
FT	Fourier transform
GNSS	Global navigation satellite system
HDPE	High-density polyethylene
HTCC	High-temperature co-fired ceramics

IC	Integrated circuit
ICI	Intercarrier interference
ICL	Iterative closest line
ICP	Iterative closest point
IDFT	Inverse discrete Fourier transform
IF	Intermediate frequency
IMU	Inertial measurement unit
IQ	In-phase-&-quadrature
ISI	Intersymbol interference
ISM	Industrial, scientific, and medical
LBE	Localized backside etching
LNA	Low noise amplifier
LO	Local oscillator
LOS	Line of sight
LRR	Long-range radar
LTCC	Low-temperature co-fired ceramics
MF	Matched filter
MIMO	Multiple input multiple output
MMIC	Monolithic microwave integrated circuit
MRR	Mid-range radar
MUSIC	MUltiple SIgnal Classification
NLOS	Non-line-of-sight
OFDM	Orthogonal frequency-division multiplexing
OGM	Occupancy grid map
OS-CFAR	Ordered statistics CFAR
PA	Power amplifier
PCB	Printed circuit board
PLL	Phase locked loop
PMCW	Phase-modulated continuous-wave
PSF	Point spread function
PSK	Phase-shift keying
PSLR	Peak-to-side lobe ratio
PTFE	Polytetrafluouroethylene
QAM	Quadrature amplitude modulation
QFN	Quad-flat no-leads
QPSK	Quadrature phase-shift keying
RANSAC	Random sample consensus
RCS	Radar cross-section
RF	Radio frequency
RMS	Root mean square
$R\text{-}v$	Range-velocity matrix
SAR	Synthetic aperture radar
SIMO	Single input multiple output
SINR	Signal-to-noise-plus-interference ratio
SISO	Single input single output

SIW	Substrate-integrated waveguide
SLAM	Simultaneous localization and mapping
SMD	Surface-mounted device
SNR	Signal-to-noise ratio
SoC	System on chip
SRR	Short-range radar
TDM	Time-division multiplexing
ULA	Uniform linear array
URA	Uniform rectangular array

Symbols

A	Amplitude
$\mathbf{a}$	Steering vector
A_q	Amplitude of the q-th target response
α	Channel attenuation
A_{eff}	Effective area of an antenna
$\mathcal{A}$	Aperture
$\mathcal{A}_{\text{SAR}}$	Synthetic aperture (for SAR)
B	Bandwidth
B_{beat}	Bandwidth of the beat signal
C	Code sequence
c_0	Speed of light (in air)
c	Speed of light (in medium)
χ	Ambiguity function
$\chi \mathbf{P}_\varrho$	Probability distribution
$\mathbf{D}$	OFDM receive signal matrix (frequency domain)
D	OFDM receive signal (frequency domain)
d	Minimal distance between two antennas of the aperture (ULAs)
$\mathbf{d}$	Vector of OFDM modulation symbols
d_{ln}	OFDM modulation symbol of the l-th symbol and n-th subcarrier
Δv	Velocity resolution
Δf	Subcarrier spacing of the OFDM subcarriers
$\Delta \psi$	Azimuth angle resolution
$\Delta \psi_{\text{SAR}}$	Azimuth resolution of the synthetic aperture (for SAR)
ΔR	Range resolution
E	Strength of the electric field
ϵ_r	Permittivity of a material
η	Index in the velocity spectrum
f	Frequency
f_c	Carrier frequency
f_D	Doppler frequency shift
$\mathcal{F}_\psi$	Field of view in angle

f_s	Sampling frequency
f_T	Transit frequency
G	Gain
I	Integration gain
$\mathcal{I}$	Radar image
K	Number of samples in fast time
k	Discrete time index
L	Number of frequency ramps in a chirp sequence (slow time)
L	Number of OFDM symbols
l	Counter/index of a chirp or signal period
Λ	Radar measurement
λ	Wavelength
m	Antenna index
m_{rx}	Receive (rx) antenna index
m_{tx}	Transmit (tx) antenna index
$\boldsymbol{\mu}_q$	Measured parameters of the q-th target (grid maps)
N	Number of OFDM subcarriers
N_s	Number of sensors
n	Counter/index of the OFDM subcarrier
n_s	Sensor index
n_0	Noise signal (time domain)
N_{ant}	Number of antennas
N_{rx}	Number of receiver (rx)
N_{tx}	Number of transmitter (tx)
N_ϱ	Number of cells
n_ϱ	Cell index
v	Index in the range spectrum
ω	Angular velocity
P	Power
p	Probability
p_{FS}	Free space (FS) probability
$\mathbf{P}_\varrho$	Cell position
p_{SNR}	SNR-based probability
p_T	Probability of a target (T)/occupancy probability
$\mathbf{P}_F^g$	Vehicle (F) position in global (g) coordinates
$\mathbf{P}_S^f$	Sensor (S) position in vehicle (f) coordinates
ϕ	Phase of a signal
ϕ_{BP}	Correction phase in the backprojection algorithms
ψ	Azimuth angle
$\varphi_{n_s}^c$	Mounting orientation of the n_s-th sensor with respect to the vehicle coordinates
$\boldsymbol{\psi}$	Matrix of transformed incident angles
ψ_ϱ	Angle under which a point is seen
Q	Number of targets
q	Counter/index of targets

Symbols

ϱ	Cell
$\mathcal{R}$	Range spectrum of the radar signal
R_ϱ	Range of a cell (to origin or sensor)
r	Radius
R	Range
$R_{\max}$	Maximum range
R_{ua}	Maximum unambiguous range
R_{FF}	Far-field range
ϱ^{g}	Cell in the global (g) coordinate system
ϱ^{s}	Cell in the local sensor (s) coordinate system
S	Slope of the frequency ramp/chirp
S	Spectral density
σ	Standard deviation
Σ	Covariance matrix
σ_ψ	Standard deviation of the azimuth angle
σ_R	Standard deviation of the range
σ_{rcs}	Radar cross-section
SNR	Signal-to-noise ratio
T	Duration of a signal period; i.e. ramp, chirp or symbol duration (depending on the modulation)
τ	Delay
T_{obs}	Observation time, total measurement duration
T_{c}	Symbol/chip duration for PMCW
T_{cp}	Cyclic prefix duration in OFDM
T_{d}	Dead time between signal periods
T_{down}	Ramp duration of a falling ramp
ϑ	Elevation angle
T_{int}	Duration of an interference
t	Time
T_{OFDM}	Total OFDM symbol duration (prefix + symbol)
T_{rep}	Ramp repetition time/interval
T_s	Sample duration/interval $1/f_s$
T_{up}	Ramp duration of a rising ramp
$\mathbf{U}$	Environment map
$\mathbf{v}_{\mathrm{c}}$	Velocity vector of the vehicle
$\mathcal{V}$	Velocity spectrum of a signal
v	Velocity
v_ω	Yaw rate of the vehicle
v_x	x-component of the vehicle velocity
v_y	y-component of the vehicle velocity
$\mathbf{v}_{\mathrm{r}}$	Radial velocity vector
v_{r}	Radial velocity
v_{ua}	Unambiguous velocity
W	Spectrum of the window function (frequency domain)
w	Window function (time domain)

w	Weight of an antenna
X	Signal (frequency domain)
x	Transmit signal (time domain)
x	Signal (time domain)
$x_{n_s}^c$	x-position of the n_s-th sensor with respect to the vehicle coordinate system
x_ϱ	x-position of a cell (in a map)
ξ	Index in the angle spectrum
$\mathbf{Y}$	Receive signal vector (frequency domain)
Y	Receive signal (frequency domain)
$\mathbf{y}$	Receive signal vector (time domain)
y	Receive signal (time domain)
$y_{n_s}^c$	y-position of the n_s-th sensor with respect to the vehicle coordinate system
y_ϱ	y-position of a cell (in a map)
y_B	Beat/IF-Signal (time domain)

List of Figures

Fig. 1.1	Radar as an active sensor principle	4
Fig. 1.2	Illustration of the radar equation	9
Fig. 1.3	Radar equation for extended targets	10
Fig. 1.4	Monostatic and bistatic radar sensors	11
Fig. 1.5	Corner reflector	12
Fig. 1.6	Illustration of the Doppler shift as elongation/compression of the emitted to the reflected wavefront due to target motion	16
Fig. 1.7	Radial velocity observed by the radar as the projection of the target velocity onto the vector between target and radar	18
Fig. 1.8	Micro Doppler signature of a person	20
Fig. 1.9	Atmospheric attenuation	22
Fig. 1.10	Scattering from rough surfaces	23
Fig. 2.1	Automotive radar with MIMO antenna system	26
Fig. 2.2	Radar sensor for level measurement at 180 GHz	29
Fig. 2.3	Body scanner	30
Fig. 2.4	Sensor for industrial quality control	31
Fig. 3.1	Discrete representation of the measurement and image space as data cubes	36
Fig. 3.2	Correlation between a complex-valued time signal and its spectrum for the example of a complex wave	39
Fig. 4.1	Block diagram of CW radar	53
Fig. 4.2	Illustration of a velocity measurement evaluation with a CW radar	54
Fig. 4.3	Block diagram of a heterodyne CW radar	55
Fig. 4.4	Realization possibilities of FMCW frequency ramps	57
Fig. 4.5	Block diagram of an FMCW radar	58
Fig. 4.6	FMCW transmit signal	58
Fig. 4.7	FMCW beat signal for the stationary case ($v \approx 0$ m/s) using a triangular signal shape	60

Fig. 4.8	FMCW beat signal with Doppler shift ($v \neq 0$ m/s) using a triangular signal shape with identical rising and falling ramp duration	62
Fig. 4.9	Illustration of the determination of the range and Doppler components with an FMCW radar	63
Fig. 4.10	Chirp sequence frequency ramps without and with additional dead time	66
Fig. 4.11	Beat signal of a chirp sequence radar for a moving target ($v \neq 0$ m/s)	68
Fig. 4.12	Spectrum of a single OFDM symbol	70
Fig. 4.13	Example of an OFDM modulation for five subcarriers	72
Fig. 4.14	Illustration of a transmit and receive sequence for a series of OFDM symbols	74
Fig. 4.15	Block diagram of a PMCW radar	77
Fig. 4.16	Binary Gold sequence of length 63 with chip rate, i.e. symbol duration T_c and alphabet $\{-1, 1\}$	78
Fig. 4.17	Autocorrelations of each three different maximum length, Gold, and Kasami sequences	79
Fig. 4.18	Cross-correlation of each three different maximum length, Gold, and Kasami sequences	80
Fig. 5.1	Two common radar signal processing chains	84
Fig. 5.2	Example of an R-v matrix: Measurement of an approaching car measured by an OFDM radar	85
Fig. 5.3	Illustration of the main lobe (beam) width and side lobe level using the example of the sinc-shaped spectrum of the rectangular window	86
Fig. 5.4	Common window functions in radar signal processing: rectangular, Von Hann, and Chebyshef windows	90
Fig. 5.5	Range measurement with a chirp sequence signal for a target at 5.3 m	99
Fig. 5.6	Range measurement with an OFMD signal for a target at 5.3 m	104
Fig. 5.7	Range measurement with a PMCW signal (maximum length sequence) for a target at 5.3 m	108
Fig. 5.8	Comparison of a constant and adaptive CFAR threshold approach	110
Fig. 5.9	Cell averaging CFAR	112
Fig. 5.10	Ordered statistics CFAR	113
Fig. 6.1	Spherical coordinates	119
Fig. 6.2	Antenna array and field superposition	120
Fig. 6.3	Schematic of a ULA and an exemplary impinging wave	124
Fig. 6.4	Uniform rectangular array	130
Fig. 7.1	Comparison of beamforming methods in case of no noise	140
Fig. 7.2	Comparison of beamforming methods	141

List of Figures

Fig. 8.1	SIMO ULA	144
Fig. 8.2	MIMO ULA	145
Fig. 8.3	Examples for the design and convolution of virtual ULAs with different transmit and receive configurations	146
Fig. 8.4	Example for the convolution of virtual apertures for a two-dimensional MIMO aperture using a two-dimensional transmit array and a one-dimensional receive array	147
Fig. 8.5	Transfer functions in a 3×3 MIMO system	148
Fig. 8.6	Transfer functions in a 1×5 SIMO system	149
Fig. 8.7	TDM-MIMO multiplexing scheme	150
Fig. 8.8	FDM-MIMO multiplexing scheme	152
Fig. 8.9	Phased Array vs. MIMO	154
Fig. 8.10	Ambiguity function of an antenna array	156
Fig. 8.11	Near-field effects in an antenna array	158
Fig. 9.1	QFN package	168
Fig. 9.2	Glass package	168
Fig. 9.3	Substrate Integrated Waveguide	169
Fig. 9.4	MMIC bonding	170
Fig. 9.5	Feeding concepts for patch antennas	171
Fig. 9.6	Gap waveguide	172
Fig. 9.7	Integrated antenna	174
Fig. 10.1	Leackage effect	179
Fig. 11.1	Exemplary visualization of a street scenario with the corresponding grid map	186
Fig. 11.2	Visualization of the three basic processing steps of grid maps	188
Fig. 11.3	Illustration of the free space model the final local map for OGMs	192
Fig. 11.4	Photo of the measurement environment with several parked cars, hedges and trees, fences, streets and gravel parking lots	194
Fig. 11.5	Visualization and comparison of AGM and OGM	195
Fig. 11.6	Representation of the vehicle coordinate system, the sensor coordinate system, the vectorial vehicle velocities, and a target	197
Fig. 11.7	Illustration of ego-motion estimation based on target lists of distributed radar sensors	198
Fig. 11.8	Illustration of the scan-matching principle for two different measurements and two different vehicle poses	199
Fig. 11.9	Sketch of a graph with eight nodes representing the vehicle poses	201
Fig. 12.1	SAR principle	204
Fig. 12.2	Visualization of the BP algorithm	209

Fig. 12.3	Amplitude and probabilistic mapping	210
Fig. 12.4	Mapping of a parking lot	211
Fig. 13.1	Interference of FMCW radars	215
Fig. 13.2	FMCW interference in frequency domain	216
Fig. 13.3	Spectrogram of an OFDM signal that is disturbed by a chirp sequence radar signal within the same frequency band	217
Fig. 13.4	Comparison of interference cancellation methods	218
Fig. 13.5	Impact of interference mitigation on an OFDM radar measurement that is disturbed by a chirp sequence interferer	220

List of Tables

Table 1.1	Important frequency bands for radar sensors in the millimeter wave range	5
Table 1.2	Typical RCS values of different targets at 77 GHz	14
Table 4.1	Example parametrization for an FMCW radar for measuring the range of a stationary target	64
Table 4.2	Example parametrization for a chirp sequence radar for measuring the range and velocity of a moving target	69
Table 4.3	Example parametrization for an OFDM radar for measuring the range and velocity of a moving target	76
Table 4.4	Example parametrization for a PMCW radar for measuring range and velocity of a moving target	81
Table 5.1	Side lobe level and main lobe width of different window functions	91
Table 5.2	Radar parameters and corresponding range resolution of the chirp sequence radar system used for the measurement in Fig. 5.5	100
Table 5.3	Radar parameters and corresponding range performance of the OFMD radar system used for the measurement in Fig. 5.6	105
Table 5.4	Radar parameters and corresponding range performance of the PMCW radar system used for the measurement in Fig. 5.7	109
Table 5.5	Decision possibilities of a detection algorithm	111
Table 9.1	Selection of some materials with permittivity ϵ_r and loss angle tan δ frequently used in the millimeter wave range of 140–220 GHz according to [[1], Table A.1]	165

Part I
Fundamentals

Chapter 1
Introduction

Radar sensor technology has spread immensely over the past decade and is now indispensable in many applications. This is mainly due to the fact that the costs of radar sensors have fallen tremendously and radar sensors have strengths in areas that are not covered by many other sensor principles. For example, radar is extremely robust against environmental influences such as weather, lighting, the position of the sun, dirt or other environmental influences in which the radar sensors are used. In addition, radar is one of very few sensor principles that allows direct velocity acquisition. Radar is an active sensor principle: A radar sensor transmits an electromagnetic wave that is reflected by a target in the radar channel and received back by the radar sensor, as shown in Fig. 1.1. By comparing the received signal with the transmitted signal, various information about the target can be obtained. The most important properties are the range of the target to the radar sensor, the radial velocity of the target in relation to the radar sensor, and the direction in which the sensor detects the target. The comparison of the received signal with the transmitted signal contributes to the robustness of the system, as interference or signals emitted by other electronic devices are not mistakenly interpreted as part of the received signal.

The principle of radar sensor technology is not new. As early as 1904, the German inventor Christian Hülsmeyer patented his "telemobiloscope", a device for detecting distant moving objects. Today, this device would be called a radar. Radar is an artificial word derived from the term *Radio Detection and Ranging*. Hülsmeyer's invention initially met with very little interest. It was only in the years before World War II that the importance of Hülsmeyer's invention became clear and military radar technology developed very quickly. For technical reasons, from today's perspective, low frequencies below the microwave range were used at that time. The use of the frequency range above 30 GHz, i.e. the millimeter wave range, was strongly driven by the vision of automated driving and automotive radar from the 1970s onwards. Today, radar sensors in the millimeter wave range are widely used in many industrial applications in automation technology, in safety technology and in the automotive sector.

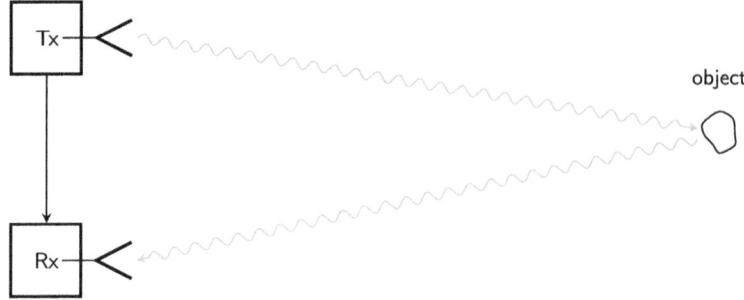

Fig. 1.1 Radar sensors are active sensors that transmit an electromagnetic wave. This wave is reflected by the target and received. By comparing the transmitted signal with the received signal, the target characteristics can be determined

In the following, the fields of application frequently used and discussed today and the corresponding frequency ranges are presented in Sect. 1.1. This is followed by the basics of radar sensor technology with a derivation of the fundamental radar equation and an introduction to key terms and measurable variables in Sects. 1.2 to 1.4. The chapter concludes with a brief insight into wave propagation in the millimeter wave range in Sect. 1.5, which includes the influence of the atmosphere and multipath propagation.

1.1 Frequency Bands and Areas of Application

The millimeter wave range is the frequency range from 30 GHz to 300 GHz. This corresponds to a wavelength from 10 mm to 1 mm. The frequency range connects upwards to the microwave range from 3 GHz to 30 GHz. The submillimeter waves follow above 300 GHz. The term *THz frequency range* frequently used today is not clearly defined; often all frequencies above 100 GHz are already referred to as THz.

Essentially, the entire frequency spectrum from direct current (DC) to well above 100 GHz is regulated worldwide, i.e. certain frequency bands may only be used for certain purposes. This prevents services or applications from interfering with each other and allows many different technical systems to coexist. However, frequency regulation is subject to constant change and is not standardized worldwide. The frequency ranges shown in Table 1.1 are often used for civil radar applications, but are only rough guidelines, which show examples of frequency ranges and bandwidths. New applications are constantly added, especially in the frequency range above 100 GHz, so that many changes are made to frequency regulation here. When designing a technical system, it is essential to research the applicable frequency regulation, i.e. the national frequency plans. In Germany, for example, the *Federal Network Agency (Bundesnetzagentur)* is responsible and has published the current frequency plan [1], which is over 700 pages long.

1.1 Frequency Bands and Areas of Application

Table 1.1 Important frequency bands for radar sensors in the millimeter wave range

Frequency range	Bandwidth	Typical applications and notes
24–24.25 GHz	250 MHz	• ISM radio band • Automotive radar (a larger bandwidth used to be available here for this application) • Industrial radar sensors (depending on the application, larger bandwidths are permitted here)
61–61.5 GHz	0.5–7 GHz	• ISM band (only 500 MHz bandwidth) • Industrial radar sensors; depending on the application and country, significantly larger bandwidths are permitted here than in the ISM band
76–77 GHz	1 GHz	• Traditional band for automotive radar, long range • Sensors for process automation
77–81 GHz	4 GHz	• Broadband automotive radar, short range • Sensors for process automation
122–123 GHz	1 GHz	• ISM radio band • Industrial radar sensors
244 GHz	2 GHz	• ISM radio band • Predominantly research

Strictly speaking, the frequency band around 24 GHz is still outside the millimeter wave range, but is often included in this range due to its importance and the similarity of the sensors used there to sensors in the millimeter wave range. This frequency band and some frequency bands further up in the spectrum are known as industrial, scientific, and medical (ISM) radio band. These are frequency bands in which electrical devices for industrial, scientific and medical purposes may generally be used in domestic areas without a license or approval, provided they comply with the general specifications of these bands. As this use is very straightforward, the ISM frequency bands are used in a large number of different applications.

The frequency range from 76 to 81 GHz is primarily used for automotive radar. While the band from 76 to 77 GHz has long been used worldwide for long-range automotive radar sensors (so-called long-range radar (LRR)), the range from 77 to 81 GHz is not yet regulated worldwide. Nevertheless, it is also approved for automotive radar in large parts of the world and is intended in particular for automotive radar applications with shorter ranges (so-called short range radar (SRR)).

The frequency bands around 122 GHz and 244 GHz can be used as classic ISM bands for many applications. Nevertheless, the limits of the currently technologically feasible radar sensors are reached here, so that at least the 244 GHz band is mainly reserved for research radars today. The first sensors or chip sets are available on the market at 122 GHz.

For many years, the trend has been towards using ever higher frequencies. This is partly due to technical feasibility and partly due to the ever-increasing absolute bandwidths available in the upper millimeter wave range. As shown in Chap. 4, large absolute bandwidths are important for radar sensors as they have a direct impact on the performance of the sensors. Furthermore, the size of most passive components, such as antennas, scales with the wavelength. For radar sensors, especially in the lower millimeter wave range, the size is often determined by the size of the antenna, so that a higher carrier frequency leads directly to smaller sensors.

1.2 Radar Equation

For the design of radar sensors and an initial assessment of whether a specific measurement task can be performed with a radar sensor, the so-called radar equation allows for a basic consideration. The radar equation relates the range of a target, the transmitted power and the characteristics of the targets to the received power.

Depending on the application, this equation must be applied and interpreted differently. In the following, a distinction is therefore made between the cases of a point target and an infinitely extended area target. In practice, however, neither case occurs exactly, so it is necessary to consider which case is closer to the specific measurement task.

1.2.1 Radar Equation for Point Targets

If the range between the target in the radar channel and the antenna is sufficiently large, the target is in the far field of the antenna and the antenna is in the far field of the target. The far field is understood to be the region of a radiated or received field in which the assumption of a local plane wave applies. The far field distance of a radiated field is given by the distance of

$$R_{\text{FF}} = \frac{2\mathcal{A}^2}{\lambda} \tag{1.1}$$

from an antenna, where $\mathcal{A}$ denotes the largest geometric length of the antenna's aperture. Although antennas are often very small in the millimeter wave range, the far-field distance R_{FF} is often only reached at a distance of a few meters from the sensor. For example, an automotive radar sensor with a frequency of 77 GHz and an aperture size of 5 cm has a far-field distance of approx. 1.3 m. A sensor for industrial applications at 122 GHz with a lens diameter of 5 cm has a far field distance of approx. 2 m. In many applications, the requirement that the sensor must also be in the far field of the target can hardly be met, as very large targets are considered in relation to the wavelength, e.g. in automotive radar or an industrial sensor for

1.2 Radar Equation

level measurements. In cases where these far field conditions can be met, the radar equation can be applied directly, as derived below. It is assumed that the target can be approximated as point-shaped, which means that the far-field condition is met.

When deriving the radar equation for point targets, it is assumed that there is exactly one target in the radar channel between the transmit and receive antennas. The transmitted power P_{tx} (the subscript tx stands for *transmit*) is radiated by the transmit antenna and is distributed isotropically on a spherical surface expanding at the speed of light c_0 around the antenna. This leads to the isotropic power density $S_{tx,i}$ of

$$S_{tx,i} = \frac{P_{tx}}{4\pi r^2}, \tag{1.2}$$

where $4\pi r^2$ corresponds to the surface of a sphere around the antenna with the radius r. Since real antennas do not radiate isotropically, but are focused with the gain G_{tx} in the main beam direction, the power density is

$$S_{tx} = \frac{P_{tx} G_{tx}}{4\pi r^2}. \tag{1.3}$$

The target is assumed to be in the main beam direction of the antenna.
The wave propagates from the transmitter to the target. This range between transmitter and target is R. The actual power density at the target is therefore

$$S_{tx} = \frac{P_{tx} G_{tx}}{4\pi R^2}. \tag{1.4}$$

This power is scattered back from the target object. The property of the target that describes this backscatter is referred to as the radar cross-section (RCS) σ_{rcs} and has the unit of an area. The RCS is discussed in more detail in Sect. 1.3.

The backscattered power is given by

$$P_s = S_{tx} \sigma_{rcs}, \tag{1.5}$$

where the subscript s stands for *scattered*. This backscattering now again leads to a wave that propagates spherically from the target in all spatial directions. This results in the power density

$$S_s = \frac{P_s}{4\pi R_s^2} = \frac{P_{tx} G_{tx}}{4\pi R^2} \sigma_{rcs} \frac{1}{4\pi R_s^2} \tag{1.6}$$

of the backscattered wave at a range R_s from the target object. If the transmitter and receiver are at the same location, the power density with $R_s = R$ at the receiver is

$$S_s = \frac{P_{tx} G_{tx}}{(4\pi R^2)^2} \sigma_{rcs}. \tag{1.7}$$

The waves with this power density are received by the receive antenna. With the effective antenna area A_{eff}, the following applies to the received power P_{rx} (the subscript rx stands for *received*):

$$P_{rx} = A_{eff,rx} S_s. \tag{1.8}$$

Adding the gain of the receive antenna

$$G_{rx} = \frac{4\pi A_{eff,rx}}{\lambda^2}, \tag{1.9}$$

the receive power becomes

$$P_{rx} = P_{tx} \frac{G_{tx} G_{rx} \lambda^2}{(4\pi)^3 R^4} \sigma_{rcs}. \tag{1.10}$$

This equation is known as the radar equation for point targets. The minimum received power $P_{rx,min}$ that can still be processed is often known for radar sensors. This allows for the maximum range R_{max} of a radar to be calculated for a given transmit power:

$$R_{max} = \sqrt[4]{P_{tx} \frac{G_{tx} G_{rx} \lambda^2}{(4\pi)^3 P_{rx,min}} \sigma_{rcs}}. \tag{1.11}$$

As can be seen, the receive power of a radar sensor decreases with the fourth power of the range R between the target and the sensor. Doubling the range leads to a reduction in receive power by a factor of 16. This is due to the two spherical waves, which propagate between the transmitter and the target with $1/R^2$ each, see Fig. 1.2, and as a product in (1.10) ultimately lead to $1/R^4$. The resulting strong attenuation of the signal in the channel between the radar sensor and the target means that a radar sensor must be able to process very weak signals. If very different ranges occur in an application, e.g. in automotive radar with targets at close range of approx. 1 m to targets up to a range of 300 m, the radar channel is said to be highly dynamic. The dynamics of the radar channel is understood as the ratio of the maximum received power from targets at close range to the minimum received power for targets at maximum range.

1.2 Radar Equation

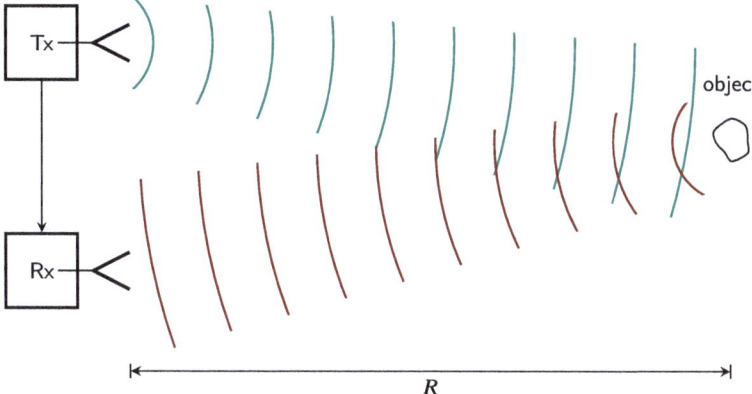

Fig. 1.2 The spherical wave emitted by the transmitter is reflected by the target (*object*) as a returning spherical wave. The range between the radar sensor and the target is greater than the far field distance

1.2.2 Radar Equation for Infinitely Extended Area Targets

For very large extended targets, the far-field condition for the targets no longer applies and no spherical wave is reflected back from the target. This is a frequently occurring case in practice, e.g. in automotive radar, which defies simple theoretical consideration. However, if it is assumed that the targets are so large that they cover the entire field of view (FoV) of the antennas, this again allows a simple theoretical description. In this case, the target is approximated by an infinitely extended target.

For the mathematical analysis of the case of an infinitely extended target, the so-called mirror method is used, see Fig. 1.3. It is assumed that the wave is reflected at the extended target object (=infinitely extended surface). This means that the transmitter and receiver are at the same location, which is equivalent to the case of a virtual receiver behind the mirror, as sketched in Fig. 1.3.

For the power density at the receiver follows

$$S_{rx} = \frac{P_{tx} G_{tx}}{4\pi (2R)^2}. \tag{1.12}$$

$2R$ is the double radar-target range resulting from the mirroring. With (1.10), the received power is

$$P_{rx} = S_{rx} A_{\text{eff},rx} = P_{tx} \frac{G_{tx} G_{rx} \lambda^2}{(4\pi)^2 (2R)^2}. \tag{1.13}$$

The maximum range is thus

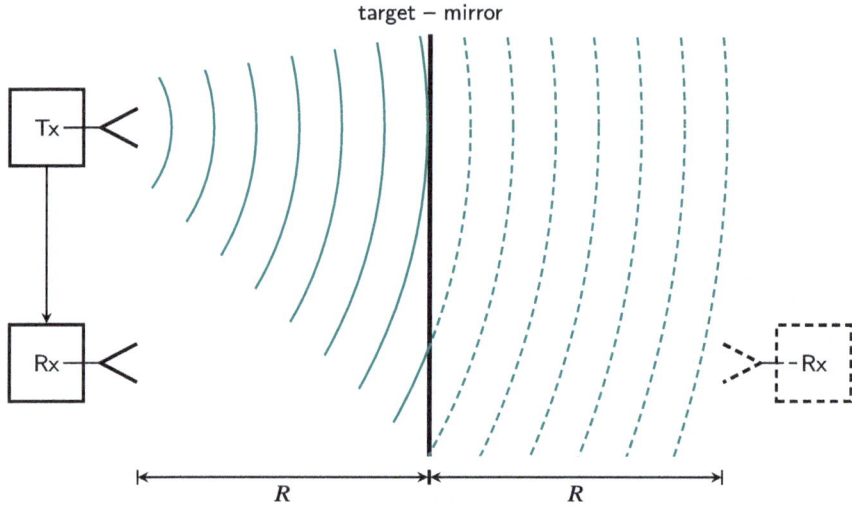

Fig. 1.3 For an infinitely extended target, the receiver is mirrored at the target according to the mirroring method, so that only one spherical wave exists

$$R_{\max} = \sqrt{P_{tx} \frac{G_{tx} G_{rx} \lambda^2}{4(4\pi)^2 P_{rx,\min}}}. \qquad (1.14)$$

Consequently, for an infinitely extended target, the received power decreases with the square of the range between the target and the radar.

Since in the millimeter wave range often neither the radar equation for point targets nor that for infinitely extended targets can be applied directly in practice, the exponent of the range dependence often has to be estimated. Depending on the application scenario, it is more likely to be 4 or 2. In some cases, the exponent is range-dependent. For example, the radar equation for extended targets applies to automotive radar at close range, while that for point targets applies at long range. The exponent changes accordingly. In addition to the line of sight (LOS), there is a second propagation path between the radar and the target in applications such as automotive radar, which influences the attenuation behavior. This is a propagation path that describes the reflection on the ground in the middle between the target and radar and is superimposed onto the LOS.

1.2.3 Monostatic and Bistatic Radars

Both monostatic and bistatic radars are used in the millimeter wave range. A monostatic radar is a radar in which the same antenna is used for transmitting and receiving. In a bistatic radar, one antenna is used for transmitting and a different one for receiv-

1.2 Radar Equation

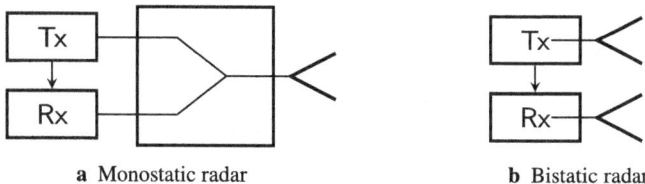

Fig. 1.4 Radar sensor with common or separate transmit (*Tx*) and receive (*Rx*) antenna. **a** In the monostatic case, the transmit and receive signals have to be separated in the radar circuit. **b** In the bistatic case, no separation is necessary due to the two different antennas

ing, as shown in Fig. 1.4. Depending on the measurement task, either monostatic or bistatic radars may have advantages. However, the differences between them can often only be recognized after an in-depth analysis of the radar sensors, as can be seen in the Chaps. 6 to 8. Obvious advantages and disadvantages are:

- Monostatic radars are smaller as only one antenna is required. This is particularly true in the lower millimeter wave range, when the size of a sensor is primarily determined by the antenna.
- Bistatic radars make it possible to optimize the antennas for their respective tasks. The transmit antenna is often used to illuminate the measurement area, while the receive antenna is used to determine the angles. As will be shown below, these requirements are often contradictory. It should be noted here that the design of modern MIMO radars, as presented in Chap. 8, can also lead to other task splits between transmitting and receiving.
- Monostatic radars require a transmit-receive splitter to separate the signals from each other. This causes losses, which is often critical, especially in the millimeter wave range.
- In the millimeter wave range, the use of bistatic radars with a high gain lens is difficult or only possible with considerable restrictions, as only one antenna can be placed in the focal point of the lens. The placement of the two bistatic antennas outside the focal point leads to a deviation of the main beam direction such that the transmit and receive antennas focus in different directions. This phenomenon, known as parallax, is particularly problematic with strongly focusing lenses.

The radar Eqs. (1.10) and (1.13) apply to the monostatic case, as only here the range R between the antenna and the target is the same on the forward and return path of the signal. In the bistatic case, the ranges can be different, which must be taken into account in the radar equations. In the millimeter wave range, the transmit and receive antennas of bistatic sensors are usually installed very close to each other, sometimes even in the same housing. Despite the bistatic design, it can then be assumed that the ranges are the same so that the radar Eqs. (1.10) and (1.13) apply.

1.3 Radar Cross-Section (RCS)

The radar cross-section (RCS) σ_{rcs} introduced in the previous section and in (1.5) is formally defined in the far field, i.e. for $R \to \infty$, as

$$\sigma_{rcs} = 4\pi R^2 \frac{|E_s|^2}{|E_i|^2}, \qquad (1.15)$$

i.e. as the ratio of the square of the magnitude of the backscattered field strength E_s to the incident field strength E_i. This equals the power ratio

$$\sigma_{rcs} = 4\pi R^2 \frac{P_s}{P_i}. \qquad (1.16)$$

The RCS corresponds to the size of a surface which, if it were to reflect isotropically, would reflect the same power as the target. Strictly speaking, the RCS is again only defined for the far field and has the unit of an area, often square meters or dBsm (Decibel square meter).

The backscatter properties of targets depend on a variety of parameters, as listed below:

Material The material that targets are made of strongly influences the RCS. Metallic objects reflect very strongly, while materials whose permittivity ϵ_r is low or similar to the permittivity of air, i.e. $\epsilon_r = 1$, reflect only weakly. Furthermore, the loss angle of materials is included in the RCS. The higher the losses of a material, the less it reflects.

Shape The geometric shape of a target is also strongly represented in the RCS. Geometries that have 90° angles cause the electromagnetic waves incident at a certain angle to be reflected back at exactly this angle. This leads to a very large RCS in relation to the geometric size of such objects.

The right-angled arrangement of three metal surfaces, as shown in Fig. 1.5, creates so-called *corner reflectors*, which have a very large RCS compared to other objects

Fig. 1.5 The corner reflector consists of three perpendicular metal surfaces. The RCS of such a structure is very large because an incoming signal is reflected in the direction of incidence. However, the reflection phase changes over the angle of incidence

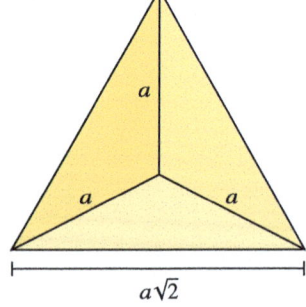

1.3 Radar Cross-Section (RCS)

of the same size. The RCS of a corner reflector can be determined from its edge length a to

$$\sigma_{\text{rcs,CR}} = \frac{4\pi a^4}{3\lambda^2}. \tag{1.17}$$

Corner reflectors are often used as defined targets in radar measurements. Similar geometries can also be found in environments where radar sensors are used. These include posts of guard rails in road traffic or metallic structures in production plants.

Targets have a particularly small RCS if all flat surfaces are aligned in such a way that no reflected waves are scattered back to the receiver.

Viewing Angle Since the reflective geometry of a target usually changes with the viewing angle, the RCS is often strongly dependent on this angle. There are only a few exceptions to this. For example, the sphere is the only object whose RCS does not depend on the viewing angle, as it is rotationally symmetrical. The RCS of the 90° geometries described above, such as the corner reflectors, also depends little on the viewing angle, as the angle of incidence is equal to the angle of reflection over a large angular range.

Size The RCS usually scales with the size of a target, provided that all other properties such as material, shape and viewing angle are the same. However, there is no uniform scaling factor with the size of a target in relation to the wavelength, as different physical effects contribute to backscattering depending on the shape of an object.

Strictly speaking, the RCS also depends on the polarization of the incident and reflected waves. A polarimetric RCS consideration leads to a matrix σ_{rcs}, which maps the co-polar and cross-polar RCS components. Furthermore, for a complete description of the RCS, a distinction must be made between a monostatic and a bistatic case. The typical case is the monostatic view, i.e. the transmitter and receiver are at the same location, so it can be assumed that the incident and reflected waves have the same or at least almost the same direction. However, if these two directions or angles are different, the RCS must be described as a function of the two angles. This is referred to as *bistatic RCS*.

Especially in the millimeter wave range, a polarimetric and bistatic description is not necessary for many applications. This is due to the fact that in this frequency range many targets are very large compared to the wavelength and therefore have many scattering centers. These many scattering centers are each dependent on a multitude of parameters, so that individual scattering centers cannot or need not be described exactly, since it is rather the superposition of the many scattering centers that is relevant. This superposition in turn depends on many parameters such as the angle of incidence, the exact reflection phase of the individual scattering centers and the exact location of the individual scattering centers. If a target produces a large number of scattering centers, it is referred to as an extended target, as opposed to a point target. In many applications, it is therefore crucial to capture the large number of reflections of an extended target as accurately and with as high resolution as possible,

Table 1.2 Typical RCS values of different targets at 77 GHz

Object	RCS in dBsm at 77 GHz
Insect	−60 to −45
Hand	−40 to −30
Pedestrian	−10 to 0
Motor vehicle	10 to 20
Trucks	20 to 40
Corner reflector	Typically many square meters, see (1.17)
Metal sphere	Circular projection surface

and not to precisely characterize a single scattering center. Finally, Table 1.2 shows some typical RCS values at 77 GHz.

If different targets with different RCS occur in an application, the ratio of the RCS of the strong targets to the RCS of the weak targets is referred to as the target dynamics. In the case of automotive radar, for example, the target dynamics between a pedestrian and a truck can be more than 40 dB according to Table 1.2. Together with the dynamics of the radar channel, the target dynamics result in the total dynamics of a radar application. This describes the ratio of the maximum to the minimum received power and is an important requirement for the hardware design of a radar sensor.

1.4 Measurands

As explained in Sect. 1.2, the electromagnetic waves transmitted by the radar are reflected by objects in the surroundings and scattered back to the radar. The waves reflected back to the radar are referred to as reflections. The amplitude of the reflection depends on the RCS of the scattering center, the so-called target, and the channel properties such as attenuation. The most important measurands that are detected with a radar are: the range, the relative radial velocity and the spatial direction of targets with respect to the radar. These measured variables are referred to below as measurement dimensions.

1.4.1 Range

Regardless of the radar type and modulation used, the signal propagation time τ of the signal trasmitted and received again at the radar is

1.4 Measurands

$$\tau = \frac{2R}{c}. \tag{1.18}$$

It depends on the range R between radar and target. The factor 2 results from the roundtrip, i.e. the path there and back, of the wave to the target at a range R; c describes the propagation speed of the wave in the respective medium. In air, it is assumed that this corresponds to the speed of light in vaccuum $c = c_0$. In modern modulations, the propagation time τ is not measured directly, but the phase or frequency shift between the emitted and received wave resulting from τ is evaluated. The exact procedure depends on the modulation and is explained in Chap. 5 for different modulation types.

1.4.2 Relative Radial Velocity

An important distinguishing feature of radars compared to other imaging sensors such as video cameras is the direct detection of the velocity of objects. Detecting e.g. the velocity of other road users plays a particularly important role in the automotive sector. A relative movement between target and radar, regardless of whether the radar itself, the target or both are in motion, results in a frequency shift of the received wave compared to the transmitted wave, which is known as a Doppler shift. This shift is illustrated as follows [2].

At time t_0, a wavefront with period T_{tx} is emitted at the radar, as shown as an example in Fig. 1.6a. At this time, there is a target at a range R_0 from the radar, which is moving radially away from it at a velocity v. The wave propagates at the speed of light c_0 and hits the target after the time $\Delta t_1 = t_1 - t_0$, which is then at the range

$$R_1 = R_0 + v\Delta t_1,$$

see Fig. 1.6c. The second period of the wave, which is transmitted at the radar at the time $t_0 + T_{tx}$ (see Fig. 1.6b), arrives after a travel time of $\Delta t_2 = t_2 - t_0 - T_{tx}$ at the target (see Fig. 1.6d), which at this point is at a range of

$$R_2 = R_0 + vT_{tx} + v\Delta t_2.$$

Both periods are scattered back to the radar and hit the receiving antenna after a total propagation time of $2\Delta t_1$ and $2\Delta t_2$, see Fig. 1.6e and f. For the range traveled and the transit time of the first period, the following therefore applies:

$$2R_1 = 2R_0 + 2v\Delta t_1 = 2c\Delta t_1 \quad \Rightarrow \quad 2\Delta t_1 = \frac{2R_0}{c-v}.$$

Similarly, the path and transit time of the second period are

$$2R_2 = 2R_0 + 2vT_{tx} + 2v\Delta t_2 = 2c\Delta t_2 \quad \Rightarrow \quad 2\Delta t_2 = \frac{2(R_0 + vT_{tx})}{c-v}.$$

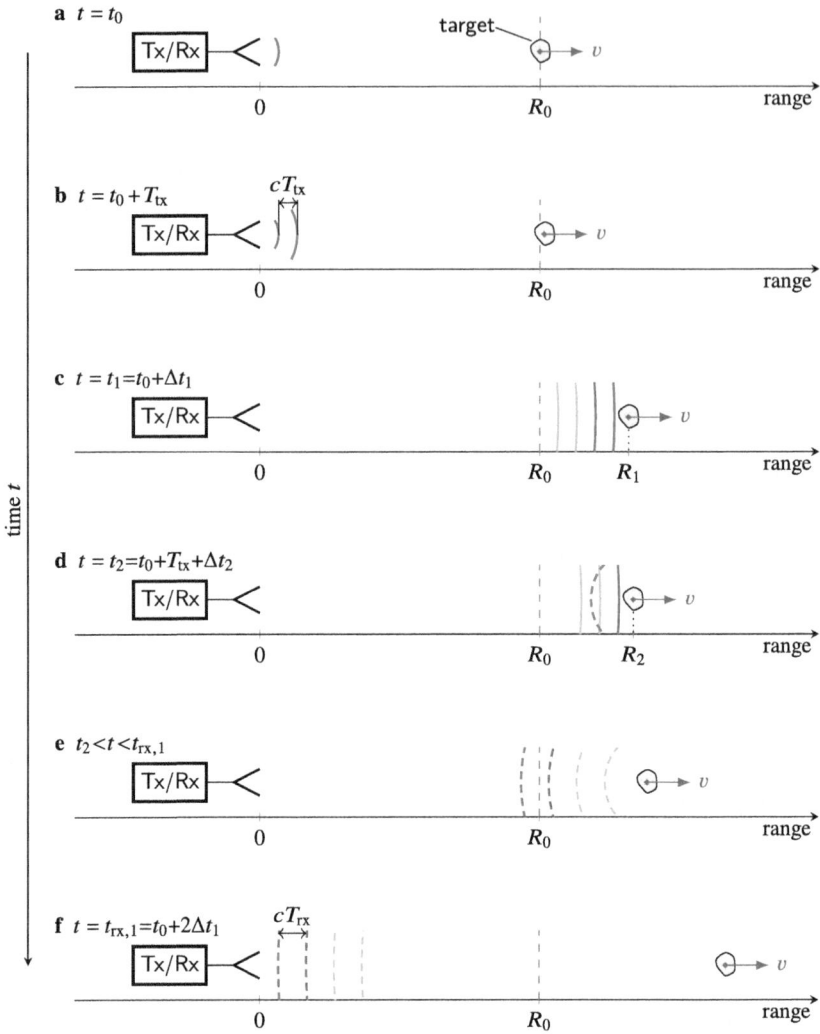

Fig. 1.6 Illustration of the Doppler shift as elongation/compression of the emitted (*solid lines*) to the reflected and received wavefront (*dashed lines*) due to the target movement using the example of a single target moving away from the sensor with velocity v. In **a** and **b**, two wavefronts with period T_{tx} are emitted by the radar at times t_0 and $t_0 + T_{tx}$. The wavefronts are reflected at the target at ranges R_1 and R_2 at times t_1 and t_2, as illustrated in **c** and **d**. After traveling back to the receiver (cfg. **e**), both waves are detected at the receiver with period T_{rx} in **f**. Due to the target motion between the arrival of the consecutive wavefronts, a change of the wavefront period happens which results in a frequency shift between transmitted and reflected (received) wave (1.19)

The first period arrives at the receiver at the time $t_{rx,1} = t_0 + 2\Delta t_1$ and the second period at the time $t_{rx,2} = t_0 + T_{tx} + 2\Delta t_2$. The difference can now be used to calculate the period length

1.4 Measurands

$$T_{rx} = t_{rx,2} - t_{rx,1} = T_{tx} + \frac{2(R_0 + vT_{tx})}{c-v} - \frac{2R_0}{c-v}$$
$$= T_{tx}\frac{c+v}{c-v}$$

of the received wave. The ratio of received to transmitted frequency is therefore

$$\frac{f_{rx}}{f_{tx}} = \frac{1-v/c}{1+v/c}.$$

In almost all radar applications, it is assumed that the velocity at which the target and radar move away from or towards each other is significantly less than the speed of light: $v \ll c$. If the denominator is replaced by the geometric series $1/(1-x) = 1 + x + x^2 + x^3 + \ldots$, where $x = -v/c$, the following approximation results:

$$\frac{f_{rx}}{f_{tx}} = \left(1-\frac{v}{c}\right)\left(1-\frac{v}{c}+\left(\frac{v}{c}\right)^2-\left(\frac{v}{c}\right)^3+\ldots\right) \approx \left(1-\frac{v}{c}\right)^2 \approx 1-\frac{2v}{c}.$$

The difference between the received and transmitted frequency is called the Doppler shift and corresponds to

$$f_D = f_{rx} - f_{tx} = -\frac{2vf_{tx}}{c} \quad \text{or} \quad v = -\frac{c}{2}\frac{f_D}{f_{tx}}. \tag{1.19}$$

This sign definition between Doppler frequency and velocity depends on the choice of reference system and can therefore also be chosen the other way round. In the following, the definition in (1.19) is used. If the radar and target are moving towards each other, $v < 0$ applies and the frequency of the received wave increases compared to that of the transmitted wave. If the two move away from each other, $v > 0$ applies and the frequency of the received wave decreases compared to the transmitted frequency. The relationship (1.19) applies to every frequency component in the signal. If the signal is modulated and consists of several frequency components, each component experiences a frequency-dependent Doppler shift. With broadband signals, this may lead to noticeable differences in the temporal distortion between the lowest and highest frequency components. However, this effect is negligible if the signal is narrowband and has a low relative bandwidth B/f_{tx} where $B \ll f_{tx}$. It should be noted that the velocity (1.19) that can be detected by the radar does not correspond to the absolute velocity v_O of the objects along their direction of movement. Rather, only the relative radial velocity $v_r = v_O \cos(\psi - \phi_r)$ of the target with respect to the radar can be detected, as shown in Fig. 1.7 as an example for three objects with different movement paths. The measurable radial velocity corresponds to the velocity components along the normal of the wavefront (LOS between target and radar) and thus to the projection of the velocity vector onto the normal vector of the wavefront at the target position.

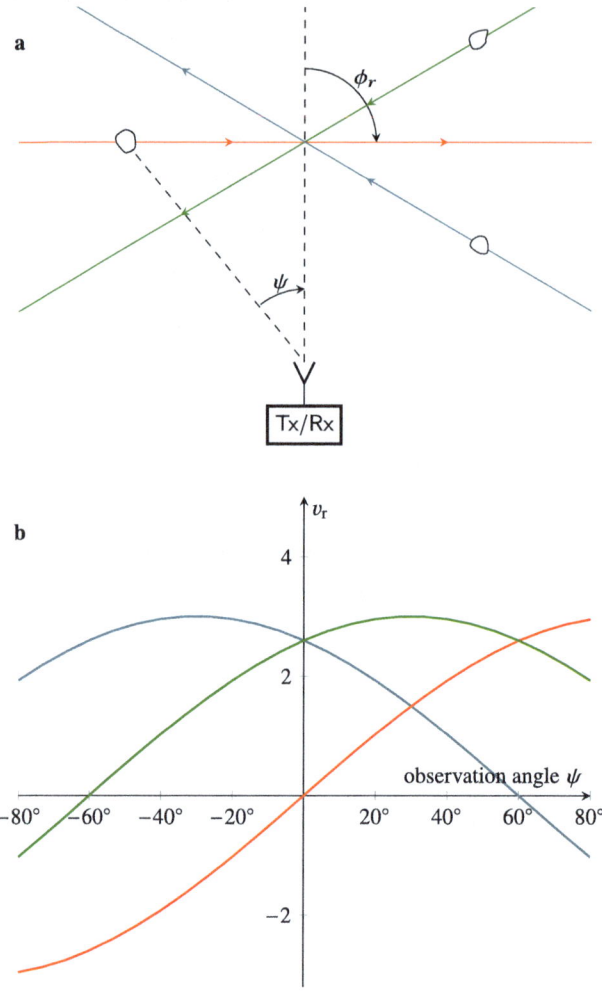

Fig. 1.7 Observed radial velocity as a function of the observation angle ψ between antenna and object and the direction of movement of the object ϕ_r. **a** three moving targets (*black*) and their lines of motion (*red, blue and green lines*). For the target moving along the *red line*, the observation angle ψ and the angle of motion ϕ_r are highlighted. **b** shows the radial velocity $v_r = v_O \cos(\psi - \phi_r)$ observed by the radar versus the observation angles for the three targets. It is assumed that the targets move at the same constant velocity v_O

In principle, the Doppler shift f_D results in a phase shift ϕ between the transmitted and received wave, which can be evaluated at the receiver. With the definition of the angular velocity

$$\omega = 2\pi f = \frac{d\phi}{dt}, \tag{1.20}$$

1.4 Measurands

the following relationship for the Doppler shift results:

$$f_D = \frac{\omega}{2\pi} = \frac{1}{2\pi}\frac{d\phi}{dt}. \tag{1.21}$$

In addition, the number of periods $\phi/2\pi$, that the wave passes through over the distance $2R$, corresponds to the ratio of the range traveled to the wavelength

$$\frac{\phi}{2\pi} = \frac{2R}{\lambda}. \tag{1.22}$$

If (1.22) is inserted into (1.21), the result is again (1.19)

$$f_D = \frac{2}{\lambda}\frac{dR}{dt} = -\frac{2}{\lambda}v = -\frac{2vf_{tx}}{c}. \tag{1.23}$$

The negative sign of the derivative of the range R is again due to the definition of the velocity relative to the sensor. The Doppler shift and thus the radial velocity is usually not measured over a single signal cycle, but over several signal cycles (signal sequences). The reason for this is that the Doppler shift f_D is often significantly lower than other signal components in the spectrum of the baseband signal. Detecting such low frequencies in the received signal requires an extremely good spectral resolution of the receiver and therefore a very long measurement duration. To increase the effective measurement duration, the Doppler shift is therefore evaluated over several signal cycles, as shown in Chap. 4.

1.4.3 Micro Doppler

Radars in the millimeter wave range make it possible to achieve extremely high resolutions in velocity measurements, which can be used in very different applications. If the Doppler shift and thus the velocity is measured with high resolution over a certain time, this results in a so-called micro-Doppler signature, as shown in Fig. 1.8.

Micro-Doppler signatures can be used very well to classify targets such as road users, gestures or animals. Micro-Doppler signatures are also used to measure vital parameters such as heartbeat and respiration.

1.4.4 Spatial Direction

In addition to detecting range and velocity, radars allow for determining the three-dimensional spatial direction of a reflection. This is made possible by mechanically or electronically changing or adapting the antenna pattern. Depending on the setup, a radar can be used to determine the *azimuth angle* in a plane parallel to the ground, the

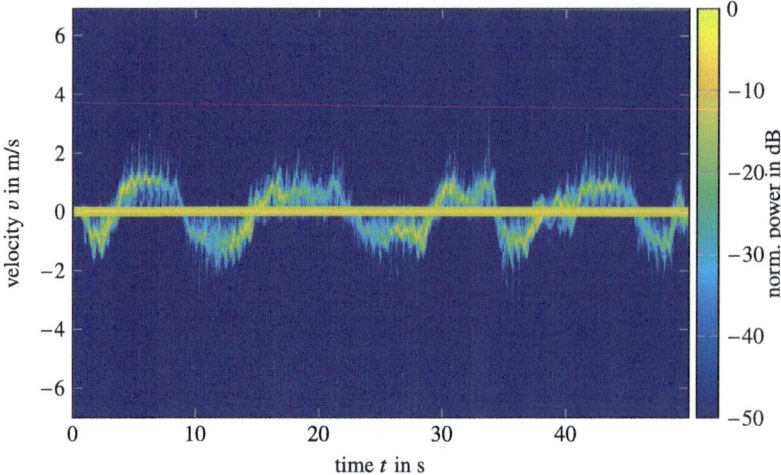

Fig. 1.8 Micro Doppler signature of a person taking approx. eight steps in one direction and then in the other. These signatures are used primarily for target classification

elevation angle in a plane orthogonal to the ground, or both angles. The procedure is identical for both planes, which is why we usually only talk about angle estimation in general. Details on this follow in the Chaps. 6 to 8.

1.4.5 Other Measurands

In addition to the usual measurands described in Sects. 1.4.1 to 1.4.4, conclusions can be drawn about other target properties such as polarization, surface properties, size, absolute velocity or direction of movement by selecting a suitable sensor configuration, waveform and signal processing. For example, the signal amplitude observed at the receiver allows conclusions to be drawn about the backscatter properties such as RCS, polarization properties or surface properties of the target. Using advanced signal processing methods such as clustering and tracking, it is also possible to estimate the size and shape of objects as well as the direction of movement, trajectory and absolute velocity.

1.5 Wave Propagation

The wave propagation of electromagnetic waves in the millimeter wave range is characterized by multipath propagation, diffraction effects and low attenuation in the atmosphere compared to the submillimeter wave range. The diffraction effects that are significantly weaker than in the microwave range, but still relevant.

1.5.1 Wave Propagation in the Atmosphere

Figure 1.9a shows the attenuation in the atmosphere versus frequency. It is noticeable that this attenuation increases with frequency beyond the upper end of the millimeter wave range and at approx. 500 GHz to over 100 dB/km. In this frequency range, radar sensors can only be used to a very limited extent for applications with longer ranges. In the millimeter wave range, the first molecular resonances occur in the spectrum, which lead to a high attenuation in the channel or absorption of the energy of the electromagnetic waves. These resonances are e.g. at 22.3 GHz, 180 GHz and 324 GHz for water vapor or approx. 61 GHz and 118 GHz for oxygen. At normal pressure, these resonances are very broad, as the quality factor of the resonance is very low due to the short free path length of the molecules. The lower the pressure, the more pronounced and narrowband the resonances become. In addition, there can be an additional attenuation of 10–15 dB/km in the millimeter wave range for violent rain as shown in Fig. 1.9b. Fog and light rain, on the other hand, do not affect the attenuation. The frequency must be selected differently depending on whether low attenuation must be avoided in order to enable long ranges or whether it is desired in order to avoid interference among different sensors. Even if the attenuation tends to increase with the frequency, it must be taken into account that higher frequencies lead to a higher antenna gain with a fixed antenna size, which in turn leads to a higher received power according to (1.10). Depending on the application and antenna choice, this can partially compensate for the larger channel attenuation.

1.5.2 Multipath Propagation

In addition to the propagation of the waves on the direct line of sight (LOS) between the target and the radar, propagation paths in which the waves are reflected, diffracted or scattered by objects in the surroundings (non-line-of-sight (NLOS)) also play a role in radar sensor technology in the millimeter wave range. All waves or signals impinging onto the receiver are superimposed vectorially, whereby the LOS always describes the shortest range and has the lowest attenuation.

Reflection from a surface is characterized by the fact that the angle of incidence and the angle of reflection are identical. In the millimeter wave range, reflection always occurs when large flat surfaces occur in comparison to the wavelength. This is often the case with tank walls, guard rails, or other large metallic surfaces. Depending on the material, the amplitude can be attenuated. However, reflections on metallic objects only lead to negligible attenuation. Transmission in materials on flat, large surfaces often plays a minor role in the millimeter wave range, as high attenuation occurs in many materials.

In the millimeter wave range, diffraction is generally understood to mean diffraction at edges. If an electromagnetic wave hits an edge, the signal is diffracted in all spatial directions from the edge. In contrast to reflection, the wave is not only emitted

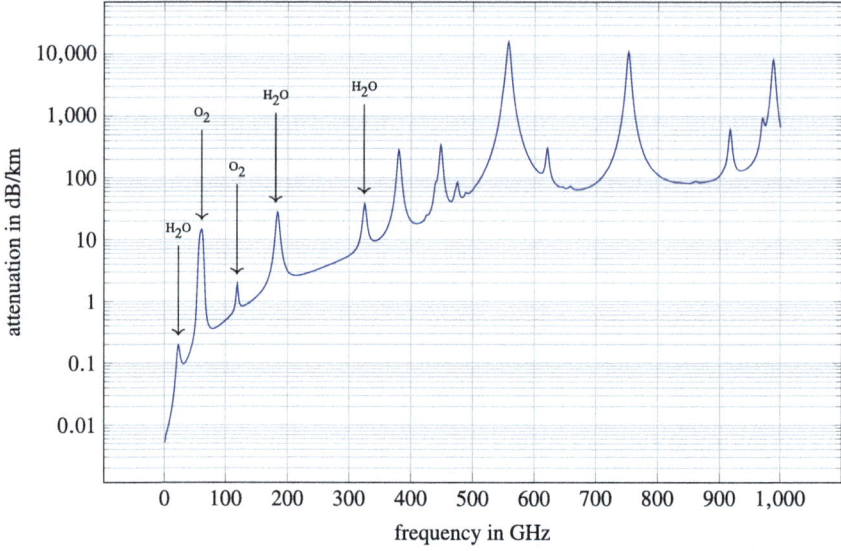

a Atmospheric attenuation in dB/km due to gases in the air

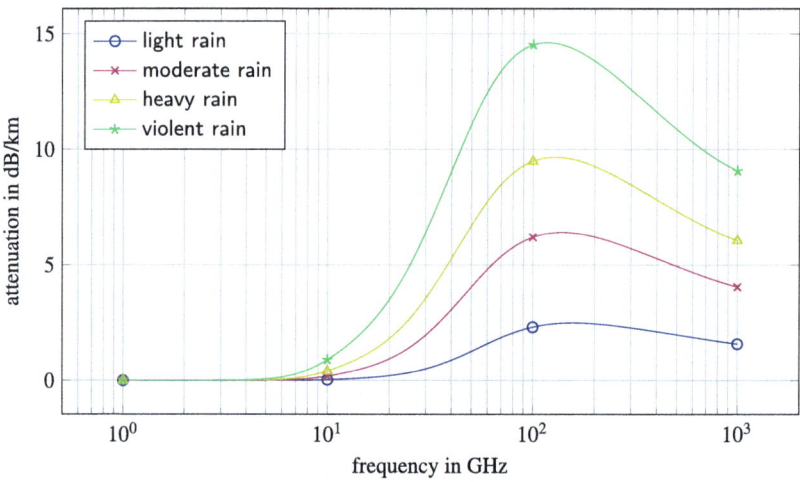

b Atmospheric attenuation in dB/km due to rain

Fig. 1.9 Atmospheric attenuation with molecular resonances and weather influences

in one direction, but in all spatial directions that are visible from the edge. As a result, a small proportion of the signal is always coming back to the receiver, regardless of where it is placed. This means that edge diffraction is an important propagation phenomenon in the millimeter wave range. The edges of targets are often clearly visible in the radar image, in contrast to metal surfaces that are not perpendicular to

1.5 Wave Propagation

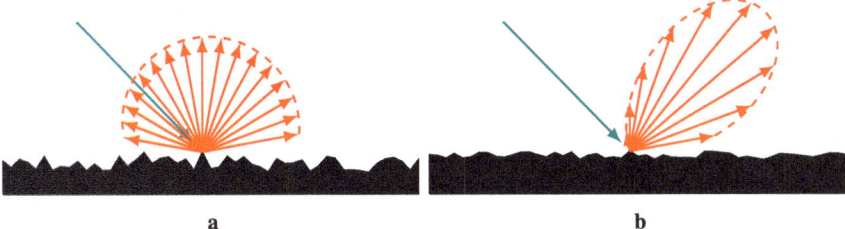

Fig. 1.10 Scattering on differently rough surfaces. In **a**, the roughness in relation to the wavelength is greater than in **b**, so that the diffuse scattering component predominates over the specular component. In **b** the roughness is lower

the LOS. Traffic signs or car door slots, for example, have edge diffraction at their edges so that they are always clearly visible in the radar image, even if the reflected signal does not hit the receiver. If the geometric dimensions of targets are smaller than the wavelength, scattering of the electromagnetic waves occurs. In the millimeter wave range, this occurs in particular with surfaces that are rough compared to the wavelength in the order of millimeters. Depending on the roughness, the diffuse, i.e. scattered in all spatial directions, or the specular signal component dominates the scattering, as shown in Fig. 1.10. As a rule, the scattering phenomenon can only be described statistically.

In the millimeter wave range, scattering often leads to unwanted signals from the channel, known as *clutter*. This occurs, for example, in automotive radar on rough road surfaces such as tar or in level measurements on undefined surfaces of bulk material.

At typical targets in the radar channel, diffraction or scattering phenomena frequently occur, which emit the wave back to the radar sensor. In practice, reflected waves generally have a much greater amplitude than diffracted or scattered waves. In everyday language, often no distinction is made as to whether the wave was reflected, diffracted or scattered at the target. Instead, the colloquial term is always reflected waves, even if this is not physically precise. Nevertheless, for the sake of simplicity, this book will also refer to reflected waves, regardless of the interaction the waves actually experienced at the target.

More detailed literature on wave propagation phenomena can be found in [3].

1.5.3 Clutter

Clutter generally refers to unwanted reflections on objects in the environment that make it difficult to detect targets. In the millimeter wave range, many structures and surfaces have a roughness that is in the order of magnitude of the wavelength. As a result, unwanted scattering phenomena often lead to clutter when signals are reflected from these rough surfaces. Examples of this are ground echoes from the road due to scattering or scattering from objects such as trees and rough house walls. Clutter generally only occurs sporadically in terms of time and space, making tracking algorithms well suited for clutter suppression.

References

1. Bundesnetzagentur, FREQUENZPLAN gemäß §54 TKG über die Aufteilung des Frequenzbereichs von 0 kHz bis 3000 GHz auf die Frequenznutzungen sowie über die Festlegungen für diese Frequenznutzungen (2022). [Online]. https://www.bundesnetzagentur.de/SharedDocs/Downloads/DE/Sachgebiete/Telekommunikation/Unternehmen_Institutionen/Frequenzen/20210114_Frequenzplan.pdf?__blob=publicationFile&v=3
2. N. Levanon, E. Mozeson, *Radar Signals* (John Wiley & Sons, Inc., 2004). https://doi.org/10.1002/0471663085
3. N. Geng, W. Wiesbeck, *Planungsmethoden für die Mobilkommunikation* (Springer, Berlin Heidelberg, 1998). ISBN 9783642589805. https://doi.org/10.1007/978-3-642-58980-5

Open Access This chapter is licensed under the terms of the Creative Commons Attribution 4.0 International License (http://creativecommons.org/licenses/by/4.0/), which permits use, sharing, adaptation, distribution and reproduction in any medium or format, as long as you give appropriate credit to the original author(s) and the source, provide a link to the Creative Commons license and indicate if changes were made.

The images or other third party material in this chapter are included in the chapter's Creative Commons license, unless indicated otherwise in a credit line to the material. If material is not included in the chapter's Creative Commons license and your intended use is not permitted by statutory regulation or exceeds the permitted use, you will need to obtain permission directly from the copyright holder.

Chapter 2
Applications

Radars in the millimeter wave range are used today in a variety of different applications that were unthinkable ten or more years ago. Radar circuits are now available highly integrated in silicon, very robust and mass-produced assembly and interconnect technologies are available and frequency regulation allows many new applications, especially in the ISM bands. The main fields of application are briefly presented in this chapter. However, as development in the field of radar sensor technology is currently progressing very rapidly, it is hardly possible to provide a complete overview of all applications.

2.1 Automotive Radar

The first attempts to integrate radar sensors into motor vehicles and use them to realize functions such as ACC (adaptive cruise control) and FCW (forward collision warning) were made as early as the 1950s. However, the millimeter wave range was only used for this purpose in the 1970s. The first radar sensors as standard equipment in motor vehicles came onto the market in 1999.

Circuits with discrete components were used in these first series sensors, e.g. Gunn elements in oscillators. The assembly and interconnect technology was also very complex and expensive from today's perspective; waveguide components were used here, for example. Compared to today, the sensors from different manufacturers differed greatly from one another. For example, different modulation methods (pulse, FSK, FMCW, see Chap. 4) were used and the antenna concepts for angle estimation were very different.

Due to the great cost pressure, a great deal has been invested in the technology in the last two decades. As a result, radar circuits at 77 GHz are now highly integrated in SiGe CMOS (complementary metal-oxide-semiconductor) and especially Si CMOS as standard. Multi-channel monolithic microwave integrated circuits (MMICs) are available that can cover the entire band from 76 to 81 GHz. Many MMICs offer three or four transmit and four receive channels, which has become a standard configuration for low-cost radar sensors. Low-cost assembly and interconnect technologies for the MMICs have also been available since around 2010, such as the eWLB (embedded wafer level ball grid array) housing [1]. Therefore, MMICs in the radar sensor no longer need to be connected with wire bonds and complex special processes are no longer required for the series production of large quantities.

Current automotive radar sensors often consist of a high-frequency circuit board on which a planar antenna array and the MMICs are integrated, as shown in Fig. 2.1. Furthermore, the sensors often contain a circuit for the power supply and a processing unit. The latest generation of MMICs already includes processing capabilities, so that all or at least large parts of the radar signal processing can be implemented there.

Today, automotive radar sensors are produced in quantities of several tens of millions per year. Due to the enormous investments in the technology and the associated cost reductions, the manufacturing costs of the sensors are in the order of magnitude of other electronic control units in vehicles.

Automotive radar sensors have become a standard sensor, partly due to their low cost and high robustness compared to other driver assistance sensors. Since the late 2010s, most vehicles in the compact class have been equipped with at least one radar sensor, the mid-range class with several sensors and up to eight sensors are often used in the luxury class. The sensors differ depending on their intended use. Front sensors are classically differentiated according to their range. Sensors with short ranges of less than approx. 50 m are called SRR, sensors with a range of 200–300 m are called

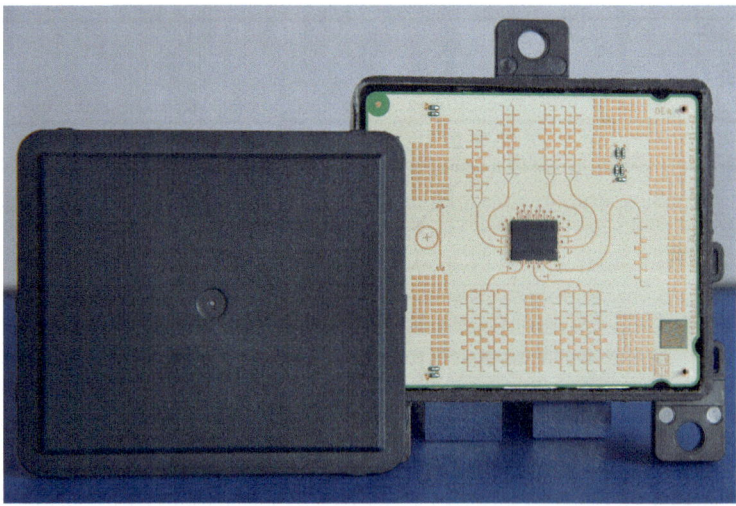

Fig. 2.1 77 GHz Automotive radar from Robert Bosch GmbH with the housing open so that the MIMO antenna system is visible. In the middle of the PCB, the IC with the radar circuitry can be seen; the distributed structures on the PCB form the MIMO antenna system

LRR and in between are the mid-range radar (MRR). Typical LRRs have a narrow field of view (FoV), as the long range is mainly necessary for high velocities, e.g. on highways. In contrast, SRRs have a very wide FoV. In addition, sensors with a wide FoV are used at the corners of the vehicle. These sensors are known as corner sensors. They have a short range due to the large opening angle. Depending on the area of application, sensors are used that are tailored to a specific function, such as sensors for blind spot detection.

Today, generally only sensors that operate in the frequency range around 77 GHz are used, as sensors at 24 GHz no longer obtain a worldwide approval. The predominant frequency band is 76–77 GHz. Chirp sequence has established itself as the modulation method that has superseded all other methods. However, different characteristics and implementation variants are realized depending on the manufacturer.

Due to the enormous demands placed on radar sensors for autonomous or fully automated driving, it can be assumed that the performance of the sensors will be further increased dramatically in the coming years. This will lead to a further differentiation in the performance of radars. Small and cost-effective sensors with just a few channels will continue to be used for classic driver assistance applications. For the new applications in the field of automated driving, it is still unclear what the most favorable sensor configurations will look like. Individual sensors with enormous performance are conceivable, e.g. with a very large number of channels, i.e. multiple input multiple output (MIMO) radars with far more than 1000 virtual channels. Or networks of simple radar sensors in which the sensors can measure in a measurement mode in which each sensor can be a receiver or transmitter for every other sensor. With such networks of simple radar sensors, it has been impressively demonstrated in recent years how the ego-motion can be estimated and thus grid maps or synthetic aperture radar (SAR) images of the environment can be created [2], see examples in Chaps. 11 and 12.

Further reading on the subject of automotive radar can be found in [3, 4].

2.2 Radars for Rail, Air and Water Vehicles

While automotive radar has been the main driver of technological progress in millimeter wave radars over the last two decades, very similar sensors have also been applied to other vehicles. For this purpose, automotive radar sensors have often been modified or equipped with customized signal processing.

Radar sensors are often used for special machinery, for example in mining. As the environmental conditions here are very harsh with heavy soiling, radar can demonstrate its robustness. The sensors are used, for example, to monitor safety zones or to measure distances to other objects. The flow rate on conveyor belts can be monitored, as can fill level measurements of mining excavator buckets or of other machines.

Radar sensors in the millimeter wave range are also used in rail transport. For example, the velocity of locomotives and wagons can be measured precisely, enabling faster shunting. Robust velocity measurement over ground is of great importance in rail traffic. For this purpose, a radar sensor is used to measure the reflection of the ground laterally or forwards and downwards. Radar sensors are also used in a similar way to driver assistance systems to monitor the surroundings of autonomous rail vehicles. On larger ships, radar sensors in the millimeter wave range are used to monitor the distances to port facilities in order to simplify the maneuvering of ships.

2.3 Sensors for Mobile Robotics

Another field of application, in which sensors similar to automotive radar sensors are often used are sensors for mobile robots such as transport robots, supply robots, etc. The sensors are used to detect the environment, as in the two applications above. In this context, radar-based grid maps of the environment are often recorded or collision avoidance is implemented. The main advantages of radars in this application are their great robustness against environmental influences and the low sensor costs.

2.4 Radars in Automation Technology

Measurement applications in automation technology are among the classic fields of application for radar sensors. While radar sensors are increasingly being used in factory automation to measure distances and velocities of objects, a typical application in process automation is level measurement.

When using industrial robots, more and more radar sensors are used to define a safety zone around the robot arm in which objects must be detected [5]. Especially in application environments where robots and humans work closely together, these monitored safety zones are essential.

Measuring fill levels in silos or tanks is an important task for radar sensors in process automation, see Fig. 2.2. The robustness of radar sensors can be ideally exploited as challenging environmental conditions often occur here, e.g. due to contamination, high pressures or high temperatures. Not only liquid levels are measured, but also the level of debris, e.g. on stockpiles. Many technologies that have emerged in recent years, driven by automotive radar, are also used in the automation area today. However, the volumes of sensors involved here are much lower, which is why concepts from assembly and interconnect technology in particular cannot always be adopted from the automotive domain directly.

Fig. 2.2 Radar sensor for level measurement in tanks and silos from Endress + Hauser SE+Co KG. Accurate measurements despite compact size: The 180 GHz sensor enables use in the smallest tanks for level measurement with an absolute bandwidth of 15 GHz. With the 3/4 inch antenna, a 6° antenna opening angle can be achieved. Image source: Endress + Hauser

2.5 Civil Security Technology

A relatively new application is body scanners for security checks, as shown in Fig. 2.3. These were first introduced at airports in test operations around 2010 and are now used very frequently. These radar systems use imaging radar sensors in the low or medium millimeter wave range to generate high-resolution images. This allows weapons to be detected under clothing, for example.

The term THz scanner, which is often used in the media, is at least misleading, as waves in the millimeter wave range are used in this case and not in the THz range. Despite the many channels required for good imaging with these sensors, the transmit power of the devices is very low—as is usual in the millimeter wave range – and is in the order of 1 mW.

Such scanners not only require a very large number of channels and high bandwidth for the resolution of the images, but also large computing power in order to implement the signal evaluation quickly despite the large number of input signals. With fast measurement and evaluation, the range of applications for the sensors is extended, for example, to any admission or access controls where large numbers of people need to be scanned.

Fig. 2.3 Body scanner R&S®QPS201 from Rohde & Schwarz at Munich Airport. The scanner at 70 - 80 GHz with more than 12000 antennas enables the detection of potentially dangerous objects hidden under clothing. Image source: Rohde & Schwarz

2.6 Quality Assurance

If millimeter waves are reflected by smooth surfaces or if these waves penetrate thin materials, inhomogeneities are often easy to detect by the reflected or transmitted field. Inhomogeneities can be small cracks, fissures, material accumulations, cavities or inclusions. For this reason, radar sensors are increasingly being used for quality assurance applications, as shown in Fig. 2.4. For example, small cracks can be detected, particularly in metallic but also dielectric materials.

A prerequisite for this type of application is that the materials or objects to be examined either reflect electromagnetic waves or can be penetrated by them. This is often the case for materials with a low water content. Optically non-transparent materials can also be penetrated, i.e. it is possible to analyze the inside of plastic parts as well as objects that are already packaged, where the radar is used to measure through the packaging.

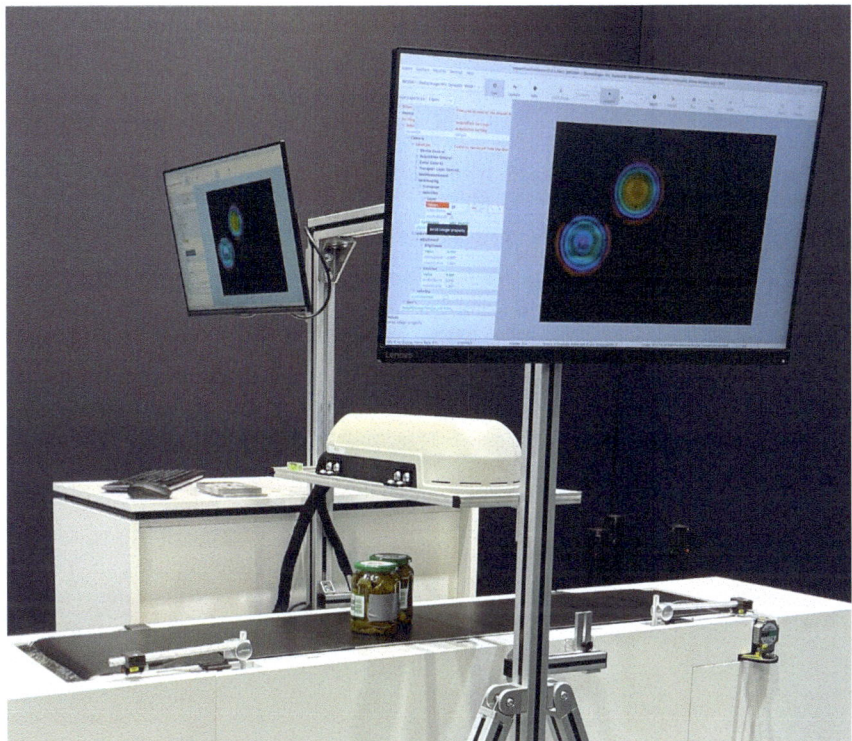

Fig. 2.4 Imaging 3D radar system from Balluff GmbH for applications in industrial quality control. The system enables all non-conductive materials to be scanned, which can be used to detect hidden objects, detect missing products, and evaluate the material properties of objects. Image source: Balluff GmbH

2.7 Medical Applications

Medical applications of radar sensors in the millimeter wave range are also relatively new or are often still at the research stage. One very intensively researched topic is the measurement of vital parameters such as heartbeat and breathing, e.g. of sleeping people. This can be used to protect infants from sudden infant death syndrome or to monitor burn victims without physical contact. The radar measures the extremely small movements of the body surface caused by heartbeat and breathing. Even the propagation of the pressure wave due to the heartbeat in the body can be observed.

In addition, a lot of research has been carried out in recent years to analyze the reflective properties of certain tissues in order to detect breast or skin cancer, for example, or to monitor the healing of wounds under bandages.

References

1. M. Brunnbauer, E. Fürgut, G. Beer, T. Meyer, Embedded Wafer level ball grid array (eWLB), in *8th Electronics Packaging Technology Conference*. (IEEE, 2006). https://doi.org/10.1109/eptc.2006.342681
2. T. Grebner, P. Schoeder, V. Janoudi, C. Waldschmidt, Radar-based mapping of the environment: occupancy grid-map versus SAR. IEEE Microw. Wireless Compon. Lett. **32**(3), 253–256 (2022). ISSN: 1558-1764. https://doi.org/10.1109/lmwc.2022.3145661
3. C. Waldschmidt, J. Hasch, W. Menzel, Automotive Radar — from first efforts to future systems. IEEE J. Microw. **1**(1), 135–148 (2021). ISSN: 2692-8388. https://doi.org/10.1109/jmw.2020.3033616
4. F. Roos, J. Bechter, C. Knill, B. Schweizer, C. Waldschmidt, Radar sensors for autonomous driving: modulation schemes and interference mitigation. IEEE Microw. Mag. **20**(9), 58–72 (2019). ISSN: 1557-9581. https://doi.org/10.1109/mmm.2019.2922120
5. M. Geiger, C. Waldschmidt, 160-GHz radar proximity sensor with distributed and flexible antennas for collaborative robots. IEEE Access **7**, 14 977–14 984 (2019), ISSN: 2169-3536. https://doi.org/10.1109/access.2019.2891909

Open Access This chapter is licensed under the terms of the Creative Commons Attribution 4.0 International License (http://creativecommons.org/licenses/by/4.0/), which permits use, sharing, adaptation, distribution and reproduction in any medium or format, as long as you give appropriate credit to the original author(s) and the source, provide a link to the Creative Commons license and indicate if changes were made.

The images or other third party material in this chapter are included in the chapter's Creative Commons license, unless indicated otherwise in a credit line to the material. If material is not included in the chapter's Creative Commons license and your intended use is not permitted by statutory regulation or exceeds the permitted use, you will need to obtain permission directly from the copyright holder.

Chapter 3
Resolution, Separability, Accuracy and Unambiguity of a Radar

Properties such as resolution, separability, accuracy, and unambiguity are important for describing the performance of a radar sensor. These properties are introduced and explained in this chapter. They form the basis for the following chapters, in which different implementation concepts for radar sensors are discussed.

In practice, the properties resolution and accuracy are often confused although they describe completely different properties of a sensor. In many applications, especially those with numerous or extended targets in the measurement scenario, resolution is the decisive parameter, while accuracy is less important. Accuracy is the decisive parameter only in application contexts where a single target is considered, such as the fill level in a tank.

Sinusoidal and periodic signals play a central role in radar signal processing and evaluation. For this reason, first some preliminary mathematical considerations are discussed in Sect. 3.1 before the respective properties are introduced in the remainder of this chapter. In Sect. 3.2, the terms resolution, separability, accuracy and unambiguity are defined and distinguished in the context of radar before they are described in detail in Sects. 3.3 to 3.6. In the following, the assumption is made that the sinusoidal and periodic signals are evaluated using discrete Fourier transforms (DFTs), as is almost always the case in practice. High-resolution spectral estimation methods represent a special case in this sense and are therefore not considered here. However, they are addressed in Chap. 7.

3.1 Preliminary Mathematical Considerations

The signal to be digitally processed by a radar sensor is either a sampled baseband time signal or a low-frequency intermediate frequency (IF) time signal. This low-frequency IF time signal is converted from a millimeter-wave carrier frequency to a comparably low frequency domain of a few MHz before sampling, for example to filter out the DC component or to reduce the $1/f$ noise influences on the signal evaluation.

The ideal sampling of a continuous-time signal $x(t)$ with sampling rate $f_s = 1/T_s$ corresponds to the multiplication of the signal with a Dirac comb as

$$x_a(t) = x(t) \sum_{k=-\infty}^{\infty} \delta(t - kT_s) = \sum_{k=-\infty}^{\infty} x(kT_s)\delta(t - kT_s) \qquad (3.1)$$

$$= \sum_{k=-\infty}^{\infty} x[k]\delta(t - kT_s). \qquad (3.2)$$

Strictly speaking, the sampling of the signal $x(t)$ corresponds precisely to the discrete points kT_s in time where $k \in \mathbb{Z}$. The temporal spacing between two samples $x[k]$ and $x[k+1]$ is referred to as the sampling interval T_s and has an influence on the performance of the radar, as explained in more detail later in this Chapter. The Fourier transform (FT) of the sampled signal $x_a(t)$ (3.1) gives its continuous spectrum

$$X_s(f) = \int_{-\infty}^{\infty} \sum_{k=-\infty}^{\infty} x[k]\delta(t - kT_s) e^{-j2\pi f t} \, dt = \sum_{k=-\infty}^{\infty} x[k] e^{-j2\pi f T_s k}. \qquad (3.3)$$

The latter representation is also referred to as discrete-time Fourier transform (DTFT). A closer look shows that the spectrum is periodic with period $f_s = 1/T_s$. Therefore follows

$$X_s(f) = \frac{1}{T_s} \sum_{k=-\infty}^{\infty} X_p(f - kf_s) \qquad (3.4)$$

where $X_p(f)$ is the spectrum of the fundamental period. The unambiguous range of the spectrum of the sampled continuous-time signal $x(t)$ is therefore determined by the sampling rate f_s respectively the sampling interval T_s. To ensure that the spectral repetitions do not overlap, the bandwidth B of the analog signal should be smaller than the reciprocal of the sample duration or the sampling frequency:

$$B < \frac{1}{T_s} = f_s. \qquad (3.5)$$

This property is the so-called Nyquist-Shannon theorem or sampling theorem.

Since both the continuous FT and the DTFT cannot be efficiently implemented and evaluated digitally, the DFT

$$X[\nu] = \sum_{k=0}^{K-1} x[k]\,\mathrm{e}^{-\mathrm{j}2\pi \frac{k\nu}{K}} \quad (3.6)$$

is usually used for digital radar signal processing of finite discrete signals. In contrast to the FT and DTFT, which both generate continuous spectra, the DFT maps a discrete-time signal of finite length to a discrete periodic frequency spectrum. Due to this property, the DFT can be implemented very efficiently on digital computers. In special cases in which K corresponds to a power of two, the fast Fourier transform (FFT) also offers a particularly fast and efficient calculation of the DFT using the Cooley-Tukey method, also known as the Radix-2 or butterfly method. This reduces the computational effort from $\mathcal{O}(K^2)$ for the DFT to $\mathcal{O}(K \log K)$ for the FFT.

The types of radar modulation considered in more detail in Chap. 4 have in common that a measurement is composed of a finite number L of a periodically repeating basic signal shape or basic period. Such a measurement consisting of L contiguous sequences is also referred to as a *frame*. Digitally, the transmit and receive frame, i.e. the discrete transmit and receive signal, is therefore often not represented as a vector but as a matrix, which contains the stacked signal periods row by row. The column number corresponds to the time sample within a period. If, in addition, a MIMO system is used, i.e. multiple transmit and receive channels as described in detail in Chap. 8, such a matrix is formed for each of these channels. The entire measurement of a frame can be visualized by stacking these matrices into a data cube, as sketched in Fig. 3.1. The dimensions of this cube are the samples $k = 0, \ldots, K-1$ of a period, often referred to as *fast time*, the periods $l = 0, \ldots, L-1$, also referred to as *slow time*, and the channels $m = 0, \ldots, N_{\mathrm{ant}} - 1$. The size of the cube is $K \times L \times N_{\mathrm{ant}}$ with $K = f_s T$ being the number of samples per signal period, L the number of repetitions, and N_{ant} the number of antennas or different channels of the radar system. As will be shown later in more detail, all these dimensions can be evaluated independently of each other in a similar way using DFTs. A DFT along the samples of a period (fast time) results in the target range, a DFT along the periods (slow time), on the other hand, results in the velocity, and finally an evaluation along the channels of the system results in the spatial angle of the target.

3.2 Introduction to Resolution, Separability, Accuracy and Unambiguity

Resolution is understood as the minimum parameter difference that two targets or reflections must have in relation to each other in order to be recognized as two distinct targets by the detector in the discrete evaluation. This difference may come about in the range, velocity or angle dimension. If the difference between two targets in relation to a parameter is worse, i.e. smaller, than the resolution of the corresponding dimension, these targets are perceived as one in this dimension, i.e. they cannot be separated from each other in this dimension. However, usually it is sufficient for

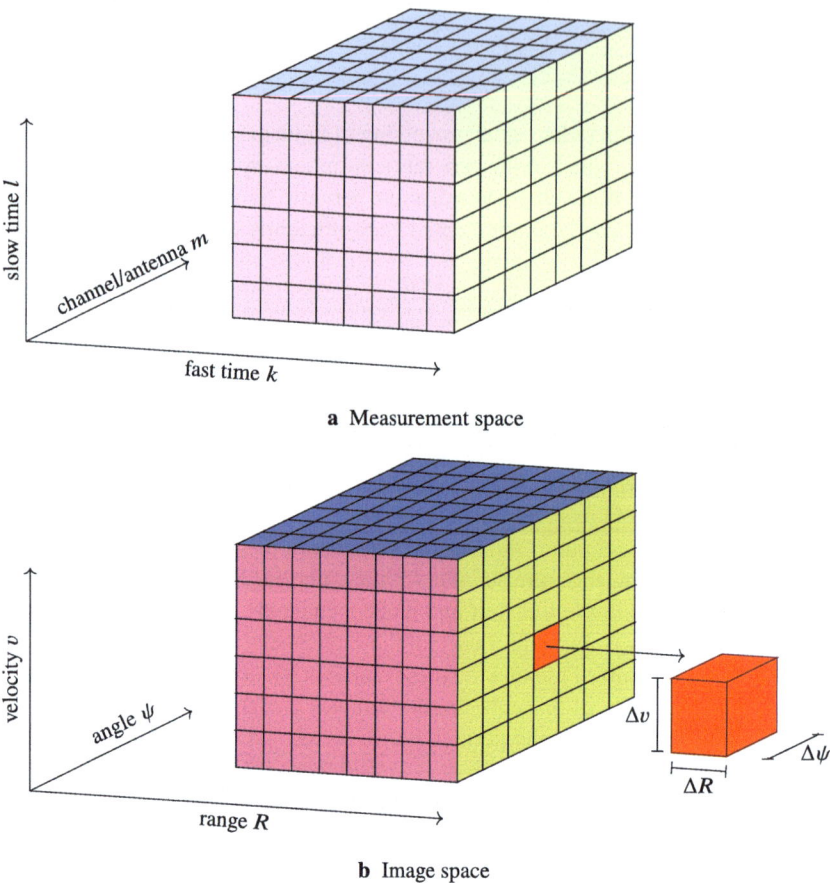

Fig. 3.1 Discrete representation of **a** the measurement space as a data cube with dimensions fast time, slow time, and channels; **b** the corresponding image space with dimensions range, velocity and angle. The cube consists of multiple resolution cubes (*red*) obtained through a multidimensional DFT

two targets to be separable in a single dimension in order to be detected as different targets. As can be seen from this definition of resolution, separability and resolution are closely related. In digital signal processing, however, this only applies in the special case of point targets that lie exactly on the discrete grid of the image area of the DFT, as is described in more detail in Sect. 3.3. The resolution is mathematically defined by the spectral resolution capability of the DFT and is thus a property of the DFT.

The separability describes the ability of the radar system to actually separate two targets from each other, taking into account the evaluation methods and algorithms used. In practice, the radar's ability to separate targets is influenced not only by the theoretical resolution, but also by properties such as the spatial extent of targets or evaluation methods used, such as windowing and the target detection algorithm (constant false alarm rate (CFAR) algorithm). An introduction to windowing and CFAR target detection is given in Chap. 5. In order to actually separate two targets from each other, a multiple of the resolution, usually 2 to 3 times, is required.

Accuracy is a measure of how accurately a measured quantity, such as the target range, can be measured under conditions such as noise; measurement errors such as quantization errors; or interference or interfering signals. These various influences can lead to errors in amplitude, phase or frequency.

Unambiguity refers to the maximum unambiguously detectable range of values of a parameter in the range, the velocity or Doppler, or in the angular dimension. Restrictions on unambiguity are imposed, among other things, by the sampling theorem, i.e., through spectral repetitions. The limit of unambiguity of a measurement dimension can usually be derived from the periodicity of the DFT.

3.3 Resolution

The resolution that a radar can achieve for a parameter such as range, velocity or angle is defined, on the one hand, by certain modulation parameters and, on the other hand, by the finite resolution capability of the DFT. The result of the signal evaluation in the various measurement dimensions can be visualized as an N-dimensional parameter space, where N corresponds to the number of measurement dimensions with associated parameters $[\theta_1, \theta_2, \ldots, \theta_N]$. The parameter space is in turn divided into a large number of discrete, N-dimensional, equally sized volumes with an extent of $\Delta\theta_1 \times \Delta\theta_2 \times \cdots \times \Delta\theta_N$, which correspond to the resolutions of the respective dimensions. Fig. 3.1b shows the typical parameter space of most radars with the three dimensions range, velocity and angle represented as a cuboid. This in turn consists of a large number of resolution cuboids of size $\Delta R \times \Delta v \times \Delta \psi$, as sketched in Fig. 3.1b, where ΔR, Δv and $\Delta \psi$ correspond to the achievable resolutions in range, velocity and angle. Assuming that all parameters are observed at Nyquist rate and no interpolation is performed, the resolution $\Delta\theta_i$ of a parameter θ_i depends on the "observation length" T_i of the signal in the respective measurement dimension, which defines the length of the DFT along this dimension. It is important to note that T_i does not necessarily correspond to a time, but can correspond to a signal bandwidth or spatial extent depending on the modulation and the parameter being observed. In general, the following applies for the resolution of the parameter θ_i:

$$\Delta\theta_i \propto \frac{1}{T_i}. \tag{3.7}$$

What this means for the parameters range, velocity and angle is described in more detail in the following.

3.3.1 Range Resolution

To determine the range resolution of a signal, it is sufficient to consider the fast-time samples of a single signal period l. This means that a single signal period of the frame is sufficient for range measurements. Regardless of the modulation used for the radar signal, the sampled received signal of a single signal period of duration $T = KT_s$ with K fast-time samples in the interval T_s is described by

$$x_l[k] = A_R\, e^{j2\pi f_R k T_s}\, \text{rect}\left(\frac{k - K/2}{K}\right). \tag{3.8}$$

Terms that do not depend on the range R or the discrete time k are summarized in the complex amplitude A_R and have no influence on the next steps. The frequency component

$$f_R = \frac{2RB}{cT} \tag{3.9}$$

is a linear function of the range R and the signal bandwidth B. This relationship applies to all modulation types considered in the following chapters. By applying a DFT (3.6) gives the ν-th coefficient of the spectrum of (3.8) to

$$X_l[\nu] = \sum_{k=0}^{K-1} x_l[k]\, e^{-j2\pi \frac{k\nu}{K}} = A_R \sum_{k=0}^{K-1} e^{j2\pi f_R k T_s}\, e^{-j2\pi \frac{k\nu}{K}} \tag{3.10a}$$

$$= A_R \sum_{k=0}^{K-1} e^{j2\pi \frac{k}{K}(f_R K T_s - \nu)} = A_R \sum_{k=0}^{K-1} e^{j2\pi \frac{k}{K}(f_R T - \nu)} \tag{3.10b}$$

$$= A_R K \operatorname{sinc}[\nu - f_R T]. \tag{3.10c}$$

The finite time signal, which is mathematically described by a rectangle, results in a sinc-shaped spectrum, which has a maximum for the coefficient $\nu = \lfloor f_R T \rceil$ ($\lfloor \cdot \rceil$: round to the nearest integer). The ν-th spectral frequency thus corresponds to $f_R[\nu] = \nu/T$. The interval between two adjacent discrete frequency components and thus the spectral resolution is therefore $\Delta f_{\text{dft}} = 1/T$. This also corresponds to the distance to the first zero of the sinc spectrum, which is also referred to as the Rayleigh bandwidth. Figure 3.2 shows the relationship between a time signal and its spectrum using a complex wave (3.8) with $f_R = 3/T$.

Now define $R[\nu]$ as the target range corresponding to the frequency component $f_R[\nu]$, and

$$R[\nu + 1] = R[\nu] + \Delta R$$

as the range that generates a frequency component

$$f_R[\nu + 1] = f_R[\nu] + \Delta f_{\text{dft}}$$

3.3 Resolution

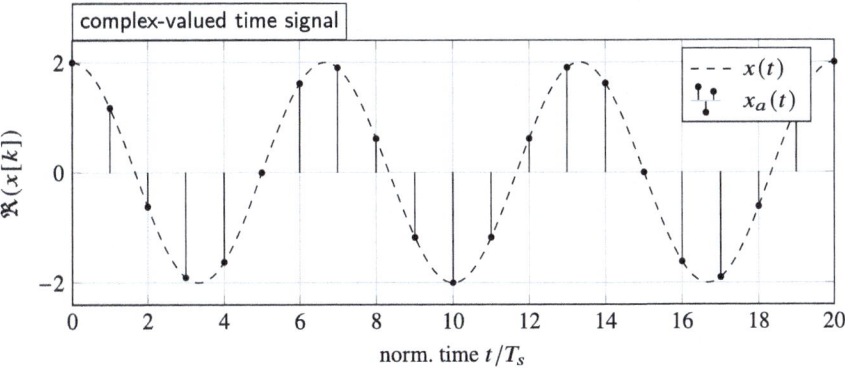

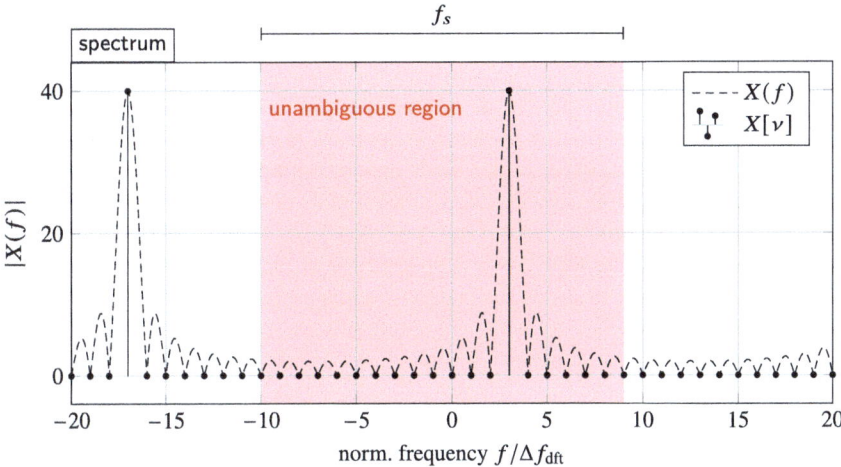

Fig. 3.2 Correlation between a complex-valued time signal (*top*) and its spectrum (*bottom*) for the example of a complex wave with frequency of $f_R = 3/T$. The sample interval of the sampled time signal $x_a(t)$ is T_s while the sample frequency spacing of the discrete spectrum is $\Delta f_{\text{dft}} = 1/T$; The spectrum follows a periodic sinc-function with period, i.e. unambiguous region (*red*) of $19/T$

that is higher by the increment Δf_{dft} compared to $f_R[\nu]$. This results in the relationship between the DFT resolution and ΔR as

$$\frac{1}{T} = \Delta f_{\text{dft}} = f_R[\nu+1] - f_R[\nu] \tag{3.11}$$

$$\stackrel{(3.9)}{=} \frac{2B(R+\Delta R)}{cT} - \frac{2BR}{cT} = \frac{2B\Delta R}{cT}. \tag{3.12}$$

Rearranging the equation gives the minimum resolvable range, i.e. the range resolution

$$\Rightarrow \Delta R = \frac{c}{2B}. \qquad (3.13)$$

The range resolution of a radar is therefore inversely proportional to the signal bandwidth B of the transmitted signal. The larger the bandwidth of a signal, the finer is the achievable range resolution, regardless of the duration of the signal or other modulation parameters. This is illustrated by the following example: To achieve a resolution of 10 cm, a signal bandwidth of about 1.5 GHz is required. For this reason, modern and future radar modulations are aiming for ever higher bandwidths. At the same time, in order to keep the relative bandwidth low, i.e. the ratio of bandwidth to the center frequency of the RF signal f_c, higher and higher carrier frequencies are being sought.

3.3.2 Velocity Resolution

While the range resolution (3.13) is determined by a DFT over the samples of a single signal period, i.e. the fast time, the velocity properties are observed over several signal periods, i.e. the slow time of the frame. To do this, it is sufficient to extract and analyze a single signal vector along any sample k of all signal periods of the frame, provided that it is the same sample number for all periods. This corresponds to the analysis of a single column of the measurement space cube in Fig. 3.1a. The discrete signal across the k-th samples in slow time is described by

$$x_k[l] = A_v \, \mathrm{e}^{\mathrm{j}2\pi f_D lT} \, \mathrm{rect}\left(\frac{l - L/2}{L}\right) \qquad (3.14)$$

and contains the Doppler frequency

$$f_D = -\frac{2 f_c v}{c}. \qquad (3.15)$$

Terms that do not depend on the velocity v or the slow time index l are summarized in the complex amplitude A_v and have no influence on the next steps. If the DFT of (3.14) is evaluated along L signal periods in the same way as (3.10), the Doppler spectrum, i.e. the velocity information

$$X_k[\eta] = \sum_{l=0}^{L-1} x_k[l] \, \mathrm{e}^{-\mathrm{j}2\pi \frac{l\eta}{L}} = A_v L \, \mathrm{sinc}[\eta - f_D L T] \qquad (3.16)$$

is obtained. Again the spectrum is sinc-shaped with a maximum at the coefficient $\eta = \lfloor f_D L T \rfloor$ and the η-th spectral frequency $f_D[\eta] = \eta/(LT)$. From this, the Doppler resolution or velocity resolution

3.3 Resolution

$$\Delta v = \frac{c}{2 f_c T L} \quad (3.17)$$

of the radar is derived analogously to the procedure for the range resolution. The velocity resolution is therefore inversely proportional to the observation time TL, i.e. the duration of a single radar measurement or frame. For example, to achieve a velocity resolution of 1 m/s at a carrier frequency of $f_c = 77$ GHz, the radar measurement must have a duration of at least 2 ms.

3.3.3 Angle Resolution

The detection of the spatial direction of incidence, i.e. the angle of incidence of the radar signal on the sensor, requires a spatial sampling of the incident wavefront. This is made possible by several transmit or receive antennas (or both) that are spatially offset by the so-called antenna spacing d. Such an arrangement is also referred to as an antenna array. The total spatial extent of the array is called the aperture. A detailed discussion of antenna arrays and multichannel radar systems can be found in the Chaps. 6 to 8. For more detailed explanations of this topic and the terms used in the following, the reader is referred to these chapters. The discussion here provides only an insight into one possibility of estimating angles using a DFT, which follows a similar principle to that of determining range and velocity from the previous sections. Beyond this, there are a number of other methods for estimating angles using radar, which are discussed in detail in Chap. 7.

The following discussion confines itself to antenna arrays in which N_ant receiving antennas are arranged at regular intervals d along a straight line. The extent or length of the array aperture is therefore $(N_\text{ant} - 1)d$. Such an aperture is referred to as a uniform linear array (ULA). With such a one-dimensional ULA, the angle of arrival, also called the direction of arrival (DoA), can be determined within the plane in which the antenna elements are located. Spatial components orthogonal to this plane cannot be determined. The regular arrangement of the antenna elements along a straight line results in a regular sampling of the phase of the incoming wavefront along the antenna array. This specific arrangement is particularly suitable for angle estimation using a DFT, similar to the DFT-based estimation for range and velocity. The angle estimation using DFT as presented below is also described in detail in Sect. 7.2.

First, the angle-dependent signal component is considered in more detail; The signal of the k-th sample of the l-th signal period can be represented as a function of the channels m of the antenna array as follows:

$$x_{kl}[m] = A_\psi \, e^{j 2\pi f_\psi m d} \, \text{rect}\left(\frac{m - N_\text{ant}/2}{N_\text{ant}} \right). \quad (3.18)$$

Here d describes the physical distance in meters between two adjacent antennas. Terms that do not depend on the angle ψ or the antenna index m are summarized in the complex amplitude A_ψ and have no influence on the next steps. The change rate of the receiving phase along the antenna positions is described by

$$f_\psi = \frac{2\cos\psi}{\lambda_c}, \quad 0 \leq \psi \leq \pi \tag{3.19}$$

which is a local oscillation in space, i.e. a function of the location, not the time; its unit is therefore $1/m$. ψ describes the angle of incidence of the wavefront on the antenna array, as shown in Fig. 6.3. A DFT applied to (3.18) results in the angular spectrum

$$X_{kl}[\xi] = \sum_{m=0}^{N_{\mathrm{ant}}-1} x_{kl}[m]\, e^{-j2\pi \frac{m\xi}{N_{\mathrm{ant}}}} = A_\psi N_{\mathrm{ant}} \operatorname{sinc}[\xi - f_\psi N_{\mathrm{ant}} d]. \tag{3.20}$$

The resulting discrete spectrum has a maximum at $\xi = \lfloor f_\psi N_{\mathrm{ant}} d \rceil$ with spatial rate of change $f_\psi[\xi] = \xi/(N_{\mathrm{ant}} d)$. By using (3.19) and rearranging the formula gives the relationship of

$$\cos(\psi(\xi)) = \frac{\xi \lambda}{2 N_{\mathrm{ant}} d} \tag{3.21}$$

between the angle ψ and the coefficients of the spectrum. The angular resolution $\Delta\psi$ thus corresponds to the angular difference between the ξ-th and $(\xi+\Delta\xi)$-th components of the spectrum, which correspond to the angles $\psi[\xi]$ and $\psi[\xi + \Delta\xi] = \psi[\xi] + \Delta\psi$, respectively. With (3.21) this results in

$$\cos(\psi[\xi] + \Delta\psi) - \cos(\psi[\xi]) = \frac{(\xi + \Delta\xi)\lambda}{2 N_{\mathrm{ant}} d} - \frac{\xi\lambda}{2 N_{\mathrm{ant}} d} = \frac{\Delta\xi \lambda}{2 N_{\mathrm{ant}} d}. \tag{3.22}$$

In contrast to the resolution in the dimensions of range and velocity, the angular resolution is usually defined as the spacing between the two zeroes closest to the maximum in the spectrum. This resolution is also referred to as Rayleigh resolution. In this case, $\Delta\xi = 2$ applies. Furthermore, due to the non-linear relationship between $\Delta\psi$ and ξ in (3.22), the angular resolution depends on the angle of arrival $\psi[\xi]$. As a result, the angular resolution is often only determined for angles in the main beam direction of the array. For the array considered here (see Fig. 6.3), this corresponds to the angle of arrival of $\psi_0 = \psi[\xi_0] = \pi/2$. Therefore (3.22) yields

$$\cos(\psi_0 + \Delta\psi) - \cos(\psi_0) = \cos\left(\frac{\pi}{2} + \Delta\psi\right) - \underbrace{\cos\left(\frac{\pi}{2}\right)}_{=0} = \frac{\lambda}{N_{\mathrm{ant}} d}. \tag{3.23}$$

This results in an angular resolution in the main beam direction of

$$\Delta\psi_{\text{ray}} = \arccos\frac{\lambda}{N_{\text{ant}}d} - \frac{\pi}{2}. \tag{3.24}$$

The angular resolution is therefore dependent on the spatial extent, i.e. the aperture $N_{\text{ant}}d$ of the antenna array. More detailed information on this can be found in Chap. 6. For example, (3.24) results in a required aperture of length 11 cm for a desired resolution of 2° at a carrier frequency of $f_c = 77$ GHz.

In some applications, the angular resolution is not necessarily defined by the Rayleigh resolution. Instead, the so-called 3dB beamwidth of the antenna array is assumed for $\Delta\psi$. The 3dB-beamwidth of the antenna beam is defined as the spacing of the two points of the beam at which the power has dropped to half of the maximum, see also Fig. 5.3. The derivation of the angular resolution for this case and further explanations can be found in Sect. 6.6.1. The resulting resolution $\Delta\psi_{\text{3dB}}$ is given by formula (6.26).

3.4 Separability

The definition of separability is closely linked to the definition of resolution. Separability is the ability of the radar to separate two different targets in one or more measurement dimensions, i.e. in range, velocity or angle. While the resolution is well defined mathematically, as shown in Sect. 3.3, and can be considered as the size of the data cube as illustrated in Fig. 3.1b, the separability is a quantity that must be considered in the context of the application. The separation capability of a radar in a particular measurement dimension is generally limited by the resolution of the system, which indicates the maximum achievable degree of separation between two targets. It is sufficient for the targets to be separable in a single measurement dimension; Consequently, it is possible to carry out measurements of several targets that are inseparable in one dimension, e.g. range, and to still separate them from one another in another dimension, for example in velocity.

The separability can be considered a property of the radar's target detector. It decides, based on defined criteria and heuristics, whether two amplitude maxima belong to two different or a single target. A common criterion is that the amplitude between two local amplitude maxima must drop by a defined factor, usually 3 dB, with respect to the lower maximum. The so-called point spread function (PSF) plays an important role in this. The PSF is the image of the impulse response of a point target by radar signal processing. This can be illustrated using the example of range evaluation by means of a DFT, as described in Sect. 3.3: The FT of an infinite continuous oscillation of frequency f_R corresponds to a spectrum $\delta(f - f_R)$. In practice, however, a time signal of finite length, i.e. a section of K samples, is evaluated instead of an infinite signal. Considering a finite section corresponds to multiplying the infinite time signal with a discrete rectangular function of the desired length of the DFT. In the spectrum, this multiplication corresponds to a convolution with the

Fourier transform of the rectangular function, i.e. with a sinc function. Instead of an infinitesimally narrow target response, this results in a PSF in the form of a sinc function with 3 dB peak width of the main maximum being about 1.2 resolution cells, as illustrated in the lower graph of Fig. 3.2.

As a consequence, to separate two equally strong targets from each other under the condition of a drop in amplitude of 3 dB, a spacing of at least 1.6 resolution cells is necessary between the targets. Certain signal evaluation methods, such as the use of a so-called windowing function to reduce the side lobe level, as discussed in Sect. 5.2, can further broaden the PSF. For this reason, the actual separation capability is usually about two to three times the DFT resolution.

3.5 Accuracy

The accuracy specifies how accurately a measured quantity such as the range, velocity or angle of a target can be measured by the radar under the influence of noise, measurement errors, interference, and other disturbances. For example, when evaluated using DFTs, the result is a perfect sinc curve with a maximum at the target range, velocity or angle (see (3.10), (3.16), (3.20)). The superposition of noise smears this ideal target response and gives rise to deviations between the estimated and the correct values. The noise process is mathematically described by a random process, which is usually assumed to be white. Consequently, the determination or estimation of the maximum of the noisy sinc curve is also described by a random process and the maximum by a random variable. By repeated measurement, its probability density distribution can be estimated and the mean measurement error determined. The variance of the distribution finally provides information about the accuracy of the measurement.

The maximum achievable accuracy that a radar system is able to achieve under the influence of noise is determined by the Cramér-Rao lower bound (CRLB). In general, the CRLB describes the minimum achievable variance of an unbiased estimator for a parameter.

The accuracy of the frequency estimation using a DFT of length $N \gg 1$ is approximately given by [1]

$$\sigma_{\hat{f}} \geq \sqrt{\frac{12}{(2\pi)^2 \mathrm{SNR} N^3 T_s^2}} \tag{3.25}$$

where T_s is the sampling interval of the time signal.

For the range accuracy, (3.9) with sample duration $T_s = T/K$ results in an accuracy of

$$\sigma_{\hat{R}} \geq \frac{Tc}{2B}\sigma_{\hat{f}_R} \stackrel{(3.25)}{=} \frac{c}{2B}\sqrt{\frac{3}{\pi^2 K \mathrm{SNR}}} \stackrel{(3.13)}{=} \frac{\sqrt{3}\Delta R}{\pi \sqrt{\mathrm{SNR}_{\mathrm{out}}}}, \tag{3.26}$$

3.5 Accuracy

where $\text{SNR}_{\text{out}} = K\,\text{SNR}$ is the signal-to-noise ratio (SNR) of a target response after the DFT. For example, given a range resolution of 10 cm, an input SNR of 20 dB and the evaluation of $K = 256$ time samples, this results in an accuracy of

$$\sigma_{\hat{R}} \geq \frac{\sqrt{3}\cdot 0.1\,\text{m}}{\pi\sqrt{256\cdot 10^{20/10}}} = 0.34\,\text{mm}\,. \tag{3.27}$$

The velocity estimation is performed by estimating the Doppler frequency f_D (3.15) by means of a DFT across several signal periods along the slow time of the signal. The sampling interval of the evaluated signal is therefore $T_s = T$, which results in an accuracy (3.25) of

$$\sigma_{\hat{f}_D} \geq \sqrt{\frac{3}{\pi^2 \text{SNR} L^3 T^2}} \overset{(3.15)}{=} \frac{2 f_c}{c}\sigma_{\hat{v}} \tag{3.28}$$

for estimating the Doppler frequency. Solved for the velocity accuracy, this results in

$$\sigma_{\hat{v}} \geq \frac{\sqrt{3}\,c}{2 f_c T L \pi \sqrt{\text{LSNR}}} = \frac{\sqrt{3}\,\Delta v}{\pi\sqrt{\text{SNR}_{\text{out}}}}\,, \tag{3.29}$$

with $\text{SNR}_{\text{out}} = L\,\text{SNR}$ being the SNR after the velocity evaluation. For example, assuming a velocity resolution of $\Delta v = 0.5\,\text{m/s} = 1.8\,\text{km/h}$, an input SNR of 20 dB and the evaluation of $L = 128$ signal periods results in an accuracy of

$$\sigma_{\hat{v}} \geq \frac{\sqrt{3}\cdot 0.5\,\text{m/s}}{\pi\cdot\sqrt{128\cdot 10^{20/10}}} = 0.002\,\text{m/s}\,. \tag{3.30}$$

The accuracy of the angle estimation is defined by the beamwidth of the antenna array and corresponds approximately to [2]

$$\sigma_{\hat{\psi}} = \frac{\psi_{3\text{dB}}}{1.6\sqrt{\text{SNR}_{\text{out}}}}\,, \tag{3.31}$$

with $\text{SNR}_{\text{out}} = N_{\text{ant}}\,\text{SNR}$ being the SNR after the angle estimation along N_{ant} channels. For example, assuming a 3 dB beamwidth of $10°$, an input SNR of 20 dB and an aperture of $N_{\text{ant}} = 16$ channels results in an accuracy of

$$\sigma_{\hat{\psi}} = \frac{8°}{1.6\sqrt{16\cdot 10^{20/10}}} = 0.125°\,. \tag{3.32}$$

In the millimeter wave range, there are only a few applications in which the accuracy of radar sensors is of great relevance. In filling level measurements in large tanks, in the measurement of piston positions or high-precision range measurements in machine tools, which are otherwise operated via glass scales, the accuracy is crucial. In typical applications of environmental monitoring, such as automotive radar sensors or sensors for robotic applications, the accuracy is often irrelevant. In these applications, the size of the resolution cells as shown in Fig. 3.1b is crucial, because only small cells allow extended targets to be mapped as accurately as possible.

3.6 Unambiguity

In the radar signal evaluation, the DFT (3.6) is employed to evaluate the discrete coefficients $X[\nu]$ of the spectrum of a signal vector $x[k]$ of finite length K. The range of unambiguity of the finite-length DFT is periodic with K due to the periodicity of the complex exponential function with 2π. Therefore, the following applies to the coefficients and the resulting spectrum:

$$X[\nu] = X[\nu + bK], \quad b \in \mathbb{Z}. \tag{3.33}$$

Thus, the maximum unambiguous index after the evaluation is $\nu = K - 1$. This property determines the unambiguous range of values when evaluating the range dimension; the same applies to velocity and angle. Eventually, the maximum unambiguous range and the associated unambiguous range of values is

$$R_{\text{ua}} = (K - 1)\Delta R \quad \text{and} \quad R \in \left[0, \; (K - 1)\Delta R\right], \tag{3.34}$$

where K is the number of unique DFT coefficients in the range evaluation.

Similarly, the maximum unambiguous velocity and the associated range of values are given by

$$v_{\text{ua}} = \left\{ -\frac{L}{2}\Delta v, \; \left(\frac{L}{2} - 1\right)\Delta v \right\} \quad \text{and} \quad v \in \left[-\frac{L}{2}\Delta v, \; \left(\frac{L}{2} - 1\right)\Delta v \right], \tag{3.35}$$

where L is the number of unique DFT coefficients in the velocity evaluation. It should be noted that different parameter ranges are defined for range and velocity in their respective spectra. The reason for this is that physically only positive ranges are possible, which is why the parameter range for the range estimation is defined as 0 to $(K-1)\Delta R$. On the other hand, both positive and negative relative velocities should be measurable when measuring velocity. For this reason, the unambiguous range for the velocity estimation is set from $-(L/2)\Delta v$ to $(L/2-1)\Delta v$. The two limits differ by a factor of Δv due to the symmetry around 0 m/s. For a sufficiently long DFT ($L > 100$) the frequently used approximation (error $\lesssim 2\%$)

$$v_{\text{ua}} \approx \pm \frac{L}{2}\Delta v \tag{3.36}$$

is sufficient. Similarly, the maximum unambiguous range is often approximated in the literature as $R_{\text{ua}} \approx K \Delta R$.

The corresponding unambiguous region for a range DFT of length $K = 20$ is illustrated in the bottom graph of Fig. 3.2. The unambiguity in the angular dimension is discussed in more detail in Chap. 6.

The unambiguous parameter range must be taken into account when designing the system, as measured targets that exceed the unambiguous parameter range may lead to ambiguities in target detection due to the periodicity of the DFT, depending on the signal modulation and filtering used. For example, it is assumed that the maximum unambiguous range of a radar with a resolution of $\Delta R = 1$ m is limited to $R_{ua} = 100$ m. If the receiver detects a strong reflection of a target at a range of $R = 105$ m, the evaluation yields a detected target at

$$R_{det} = R \bmod (R_{ua} + \Delta R) = 4 \text{ m}.$$

As a possible consequence, an emergency braking system would detect an extremely close object and would trigger an emergency braking maneuver, even though there was no danger. For this reason, it is important to choose a sufficiently large unambiguous range in the system design so that either reflections outside the chosen unambiguous range do not occur or are so weak that they are not picked up by the detector.

References

1. S. Scherr, R. Afroz, S. Ayhan et al., Influence of radar targets on the accuracy of FMCW radar distance measurements. IEEE Trans. Microw. Theory Tech. **65**(10), 3640–3647 (2017). https://doi.org/10.1109/tmtt.2017.2741961
2. G. Curry, *Radar System Performance Modeling* (Artech House, 2005). ISBN: 9781580538169

Open Access This chapter is licensed under the terms of the Creative Commons Attribution 4.0 International License (http://creativecommons.org/licenses/by/4.0/), which permits use, sharing, adaptation, distribution and reproduction in any medium or format, as long as you give appropriate credit to the original author(s) and the source, provide a link to the Creative Commons license and indicate if changes were made.

The images or other third party material in this chapter are included in the chapter's Creative Commons license, unless indicated otherwise in a credit line to the material. If material is not included in the chapter's Creative Commons license and your intended use is not permitted by statutory regulation or exceeds the permitted use, you will need to obtain permission directly from the copyright holder.

Part II
Radar Types, Modulation Schemes and Radar Imaging

Chapter 4
Radar Types and Modulation Schemes

As explained in Chap. 3, radar sensors are used, among other things, to measure ranges and velocities. According to the radar equation (1.10), the amplitude of a radar receive signal appears to be proportional to $1/R^4$, which should enable a range measurement to be carried out by determining the received signal level. In practice, however, many effects affect the received signal level, making it impossible to use the signal amplitude for range measurements. In particular, multipath propagation, atmospheric effects, beam-dependent antenna characteristics, the properties of extended targets and, ultimately, the properties of the receiver circuit all influence the amplitude of the received signal, so that it is not exclusively dependent on the target range.

Instead, to determine the target range, the transmitted signal is modulated. There are many different modulation methods known for this purpose. The classic pulse modulation, in which short pulses are transmitted in the time domain, was used frequently in the past but is now rarely used in the millimeter wave range because this method cannot be implemented cost-effectively in modern radar circuits. Instead, methods in which transmission and receiving are continuous, so-called continuous wave (CW) methods, have become established. These are presented in the Sect. 4.1 on analog modulation methods. With these methods, the measurement of the radial target velocity is also taken into account: this is always achieved by evaluating the Doppler frequency shift between the transmitted and received signals.

Since the mid-2010s, the use of digital modulation techniques for millimeter-wave radars has been widely discussed, at least in the research community. These radars offer enormous flexibility in the modulation design, but also require the processing of enormous amounts of data. Since it is becoming apparent that these radars are likely to replace analog-modulated radars at least in part in the medium run, they are presented in Sect. 4.2.

© The Author(s) 2025
C. Waldschmidt et al., *Millimeter Wave Radar*,
https://doi.org/10.1007/978-3-031-89118-2_4

4.1 Analog Modulation Schemes

Today, almost exclusively analog CW modulation methods are used for millimeter-wave radars. Due to their continuous transmission of the transmit signal, only low transmit power levels are required to fulfill the requirements regarding the radar equation. This means that radar circuits can be easily implemented in highly integrated silicon technologies, which has led to extremely cost-effective sensors and thus to the widespread use of radar sensors in the millimeter wave range. Compared to pulse-based radars with very long ranges, CW radars have the disadvantage that the return signal must be received while the transmit signal is being transmitted. Since in compact radars only a finite degree of decoupling between the transmitter and receiver is possible, the hardware of CW sensors must be designed in such a way that weak return signals can be evaluated in the receiver despite a strong coupling of the transmitted signal. This issue is examined in more detail in Chap. 9.

A major advantage of analog radar sensors, in particular FMCW and its derivative chirp sequence radars, is that the sensors' wide bandwidth in the millimeter wave range is mapped onto a relatively narrow bandwidth in the baseband, i.e. the conversion rates at the transition to the digital domain are much lower than the sensors' radio frequency (RF) bandwidths. Furthermore, for these sensors, the frequency of the baseband signal is proportional to the target range. Since, according to the radar equation, the level of the received signal decreases with the fourth power of the range, the level of the received signal also decreases with frequency with the fourth power of the range. This can be compensated for in the hardware design by a highpass filter in the baseband, so that the reception dynamics of FMCW-based sensors are significantly smaller than those given by the radar equation. The following section will first introduce non-frequency-modulated CW radar, followed by the currently predominant modulations, FMCW and chirp sequence.

4.1.1 CW Radars

For a measurement, CW radars transmit an unmodulated continuous wave

$$x(t) = A \cos\left(2\pi f_c t + \phi_0\right) \qquad (4.1)$$

with a constant frequency f_c. Due to the continuous nature of the signal, the radar's transmit power is constant and no high spikes occur in the transmit power during the measurement. The CW radar is particularly suitable for measuring velocity, which is why it is often referred to as a Doppler radar. The basic architecture of a simple CW radar is depicted in Fig. 4.1. The single-frequency local oscillator (LO) signal is fed to the antenna via a circulator. In most cases, the signal is first amplified by a power amplifier (PA). The antenna emits the signal into the transmit channel, where it is reflected by targets. After receiving the reflected signal, the circulator forwards the

4.1 Analog Modulation Schemes

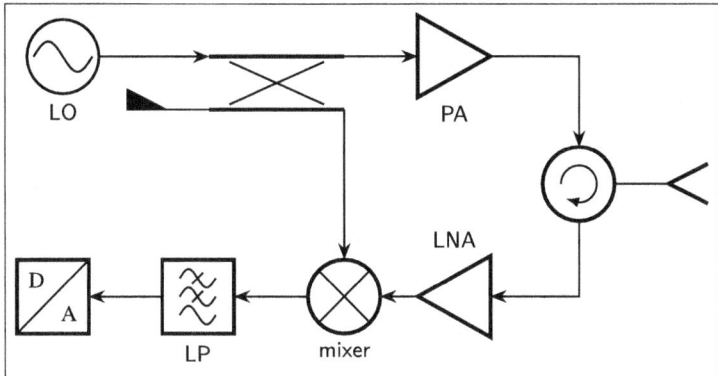

Fig. 4.1 Block diagram of a homodyne CW radar. The LO generates a sinusoidal signal that is radiated via the antenna after amplification in the power amplifier. The receive signal is amplified in an LNA and then mixed with the transmit signal. The beat signal, whose frequency results from the difference between the transmit and receive signals, is present at the output of the mixer

signal to the receiver. Depending on the requirements for target dynamics and maximum power, the signal is first amplified in the receiver using an low noise amplifier (LNA). However, in the millimeter wave frequency range, very high received signal levels occur in some applications, making it impossible to design a useful LNA. In this case, it is omitted. At the mixer, the received signal is mixed with the transmitted signal, resulting in the so-called beat signal at the difference frequency of the transmitted and received signals. In general, the received signal $y(t)$ of the radar corresponds to the transmitted signal $x(t)$ delayed by the time τ and attenuated by the channel attenuation factor α:

$$y(t) = \alpha x(t - \tau). \tag{4.2}$$

The channel delay

$$\tau = \frac{2R}{c} + \frac{2vt}{c} = \frac{2R}{c} - \frac{f_D}{f_c} t \tag{4.3}$$

is proportional to the target range R and its velocity v relative to the radar, as detailed in Sect. 1.4. The upper graph in Fig. 4.2 shows an example of a CW transmit and receive signal for a single target with constant velocity. The time delay and the frequency shift between the two signals are clearly recognizable. If (4.3) is inserted into (4.2), the received signal yields

$$y(t) = \alpha A \cos\left(2\pi f_c t - 2\pi f_c \tau\right) \tag{4.4}$$

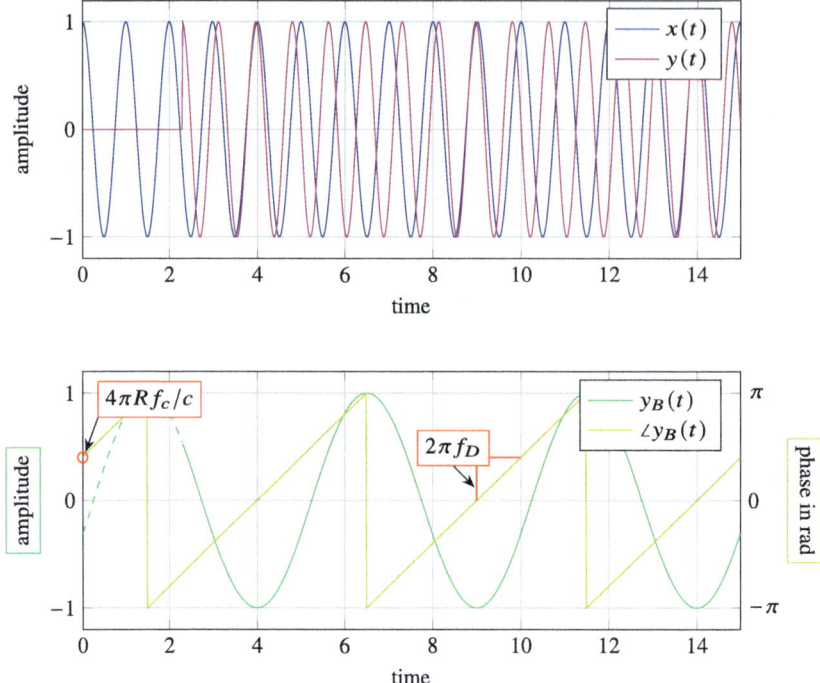

Fig. 4.2 Illustration of a velocity measurement evaluation with a CW radar. The *upper graph* shows the transmitted CW signal (*blue*) and the received signal (*red*) reflected from a target moving at velocity v. The time axis is normalized to the start of transmission, therefore, the receive signal is zero at the beginning until the reflected wave reaches the receiver (time of flight). Due to the target motion, the wave experienced a Doppler frequency shift, therefore, the *red* wave is compressed compared to the *blue* one. The *lower graph* shows the amplitude (*green*) and phase (*yellow*) of the resulting beat signal (4.6). The starting phase $4\pi R f_c/c$ depends on the delay, i.e. range while the slope of the phase $2\pi f_D$ depends on the Doppler frequency

$$= A_{\text{rx}} \cos\left(2\pi\left(f_c t - \frac{2R f_c}{c} + f_D t\right)\right). \qquad (4.5)$$

Mixing the transmit and receive signals and then filtering them through a lowpass filter gives the baseband or beat signal

$$y_B(t) = A_{\text{rx}} \cos\left(2\pi\left(f_D t - \frac{2R f_c}{c}\right)\right). \qquad (4.6)$$

When measuring moving targets, the baseband signal $y_B(t)$ is therefore an oscillation with the Doppler frequency $2\pi f_D$, since the received signal experiences a Doppler shift of the transmitted frequency proportional to the velocity of the target. The range, on the other hand, is only reflected in the initial phase $4\pi R f_c/c$.

4.1 Analog Modulation Schemes

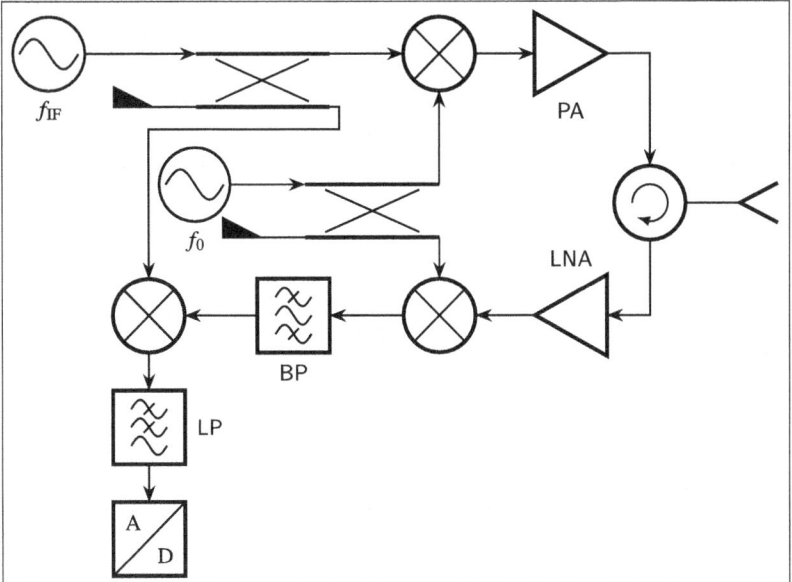

Fig. 4.3 Block diagram of a heterodyne CW radar. In the heterodyne configuration, two LO signals are used to realize an intermediate frequency

The lower diagram in Fig. 4.2 shows the amplitude and phase of the beat signal (4.6), which correspond to the transmitted and received signals in the upper graph. The range-dependent initial phase and the velocity-dependent slope of the phase progression can be extracted directly from this.

However, the simple homodyne architecture of a CW-radar as shown in Fig. 4.1 has several disadvantages. Firstly, very small Doppler frequencies are strongly affected by $1/f$ noise and are therefore difficult to detect. Secondly, in the baseband spectrum, negative Doppler shifts cannot be distinguished from positive ones, since negative frequencies of the lower sideband are mirrored into the upper sideband. Consequently, the sign of the Doppler shift cannot be determined, which means that no statement about the direction of motion is possible. The use of a heterodyne architecture as shown in Fig. 4.3 allows the received signal to be mixed into a low IF frequency range instead of directly into the baseband. This gives a signal at the frequency

$$f = f_{\text{IF}} - f_D, \tag{4.7}$$

which allows the sign of the Doppler shift to be evaluated around f_{IF}.

The CW radar is only suitable for range measurement to a limited extent. The range of stationary objects ($v = 0$ m/s) can be determined by evaluating the phase of the received signal relative to the transmitted signal using (4.6) with $f_D = 0$ Hz via

$$y_B(t) = A_{\text{rx}} \cos\left(-2\pi \frac{2R f_c}{c}\right). \tag{4.8}$$

However, due to the periodic nature of the phase, this range measurement is limited to ranges up to

$$R_{\text{ua}} = \frac{\lambda_c}{2}, \tag{4.9}$$

which, for example, already limits the unambiguous range to $R_{\text{ua}} = 15$ cm at $f_c = 1$ GHz. Range measurements are therefore only possible at extremely close ranges or at very low frequencies.

A simple way to increase the unambiguous range is to use different CW signals of different frequencies, which are transmitted in succession. The method is reminiscent of frequency-shift keying (FSK) in communications engineering, which is why this modulation is called FSK radar. Comparing the received phases of two CW signals of different frequencies yields an unambiguous range of $\frac{\lambda_\Delta}{2} > \frac{\lambda_0}{2}$. In this case, $\lambda_\Delta = c/|f_1 - f_2|$ corresponds to the wavelength of the difference frequency of the two individual CW signals, which is always greater than the wavelengths of the two individual CW signals. However, if there are multiple targets in the channel, ambiguities arise that may be resolved by increasing the number of CW signals of different frequencies. Since the number of frequencies required increases rapidly with the number of targets, FSK radar is rarely used in practice. Instead, frequency-modulated CW signals are commonly used, a technique that is described in the next section.

4.1.2 FMCW Radars

If a frequency-modulated signal is transmitted instead of a single-frequency or monotone signal, the method is referred to as frequency-modulated continuous-wave (FMCW) radar. Usually, a linear frequency modulation is chosen, in which the transmit frequency is increased or decreased linearly over time, thus generating a so-called frequency ramp or chirp. The instantaneous transmit frequency

$$f(t) = f_c \pm St \tag{4.10}$$

is thus a function of time with a starting frequency f_c and a ramp slope

$$S = \frac{B}{T}, \tag{4.11}$$

which is determined by the desired frequency range or bandwidth B of the ramp of finite duration T. The FMCW signal consists of a series of L frequency ramps. There are a number of different ramp shapes that can be used. Depending on the sign

4.1 Analog Modulation Schemes

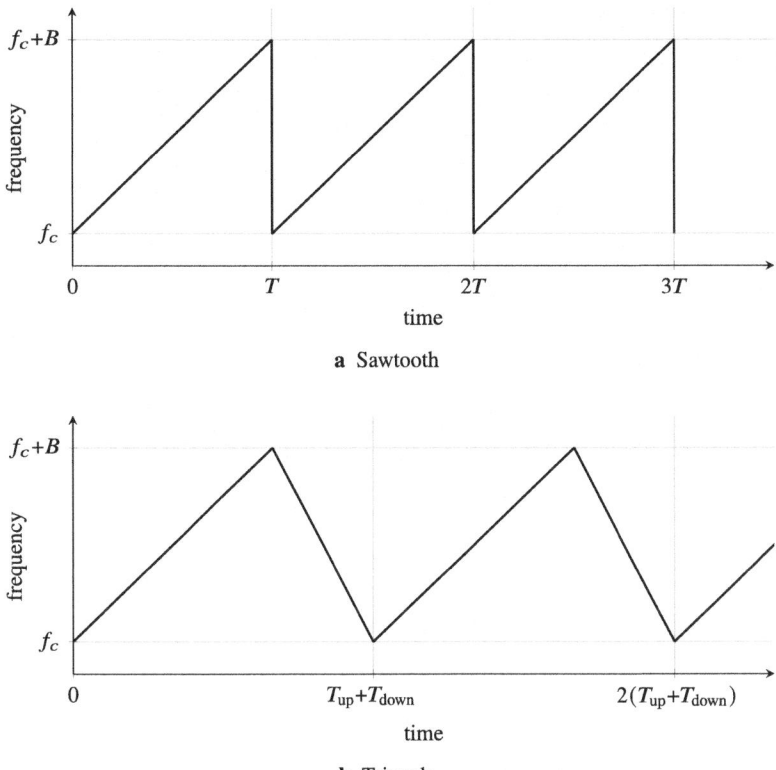

Fig. 4.4 Realization possibilities of FMCW frequency ramps: **a** the sawtooth pattern only consists of rising slopes while **b** the triangle pattern has both rising and falling slopes. The rising and falling slopes of the triangle must not be identical

selected in (4.10), the ramps can be falling (−) or rising (+) relative to the starting frequency; the latter gives the so-called sawtooth pattern as shown in Fig. 4.4a. A rising and a falling ramp can be combined to form a triangular ramp, as shown in Fig. 4.4b. In this case, the slopes of the rising and falling ramps may differ.

Like CW radars, FMCW radars feature a constant transmit power without power spikes. A block diagram of an FMCW radar is shown in Fig. 4.5, with the only difference to a CW radar being the signal source. The frequency modulation, i.e. the ramp generation, is usually implemented using a phase locked loop (PLL), [1–3].

The real-valued transmit signal of a single time-dependent frequency ramp or modulation period is modeled in the time domain with

$$x_0(t) = A \cos\left(\phi_{tx}(t)\right), \quad 0 \leq t < T. \tag{4.12}$$

For the sake of simplicity, the following mathematical description is based on a single rising or falling frequency ramp, which is expressed in (4.13) by a negative (falling)

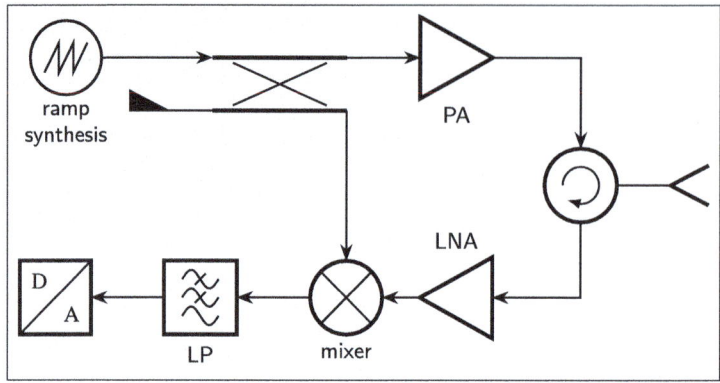

Fig. 4.5 Block diagram of an FMCW radar. The block diagram is similar to that of a CW radar, only the signal source is different and must be able to generate frequency ramps

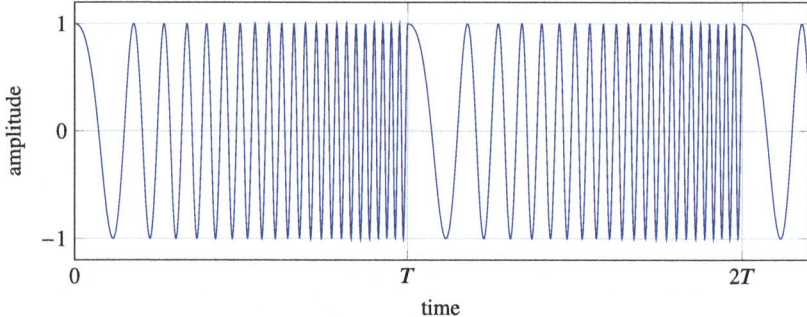

Fig. 4.6 FMCW transmit signal with period T in the time domain

or positive (rising) sign, respectively. The phase $\phi_{tx}(t)$ of a single ramp (4.12) is given by the integral over the frequency $f(t)$ (4.10):

$$\begin{aligned}\phi_{tx}(t) &= 2\pi \int_0^t f(t) dt = 2\pi \int_0^t \left(f_c \pm \frac{B}{T}t\right) dt \\ &= 2\pi \left(f_c t \pm \frac{1}{2}\frac{B}{T}t^2\right) - \phi_0.\end{aligned} \qquad (4.13)$$

Thus, the time signal of a single FMCW frequency ramp (4.12) is

$$x_0(t) = A\cos\left(2\pi\left(f_c t \pm \frac{1}{2}\frac{B}{T}t^2\right) - \phi_0\right). \qquad (4.14)$$

A series of frequency ramps, as shown in Fig. 4.6, finally forms an FMCW radar signal or FMCW frame.

4.1 Analog Modulation Schemes

The FMCW receive signal is shifted by the delay τ (4.3) with respect to the transmit signal (4.14). For a single frequency ramp, this results in the received signal

$$y(t) = x_0(t - \tau) = A \cos\left(2\pi\left(f_c(t - \tau) \pm \frac{1}{2}\frac{B}{T}(t - \tau)^2\right) - \phi_0\right) \quad (4.15)$$

$$= A \cos\left(\phi_{rx}(t)\right) \quad (4.16)$$

where the received phase is

$$\phi_{rx}(t) = \underbrace{2\pi\left(f_c t \pm \frac{1}{2}\frac{B}{T}t^2\right) - \phi_0}_{\phi_{tx}(t) \to (4.14)} - 2\pi\left(f_c \tau \mp \frac{B}{T}t\tau \pm \frac{1}{2}\frac{B}{T}\tau^2\right). \quad (4.17)$$

At the receiver, the signal (4.16) is first mixed with the transmitted waveform (4.14) using an in-phase-&-quadrature (IQ) mixer and then sampled; this gives the complex-valued beat signal

$$y_B(t) = y(t)x^*(t) = A e^{j\phi_B(t)}. \quad (4.18)$$

The phase $\phi_B(t)$ of the beat signal is derived mathematically from the difference between the phases of the transmitted (4.13) and received signals (4.17)

$$\phi_B(t) = \phi_{tx}(t) - \phi_{rx}(t) = 2\pi\left(f_c \tau \pm \frac{B}{T}t\tau \mp \underbrace{\frac{1}{2}\frac{B}{T}\tau^2}_{\ll}\right). \quad (4.19)$$

The term $B/(2T)\tau^2$ can usually be neglected, as it is significantly smaller than the other factors.

4.1.2.1 Range Measurements with FMCW Radars

First, the range measurement using an FMCW radar is considered for stationary targets or targets with a low relative velocity. Under the condition $v \approx 0$ m/s, the target-dependent channel delay τ (4.3) results in $\tau \approx 2R/c$. If this is substituted into (4.19), the phase of the beat signal becomes

$$\phi_B(t) \approx 2\pi\left(f_c \frac{2R}{c} \pm \frac{B}{T}\frac{2R}{c}t\right). \quad (4.20)$$

Differentiating the phase (4.20) with respect to time t yields the beat frequency of a rising or falling ramp for the stationary case ($v \approx 0$ m/s)

$$f_B = \frac{1}{2\pi}\frac{d\phi_B(t)}{dt} = \pm\frac{2B}{T}R. \quad (4.21)$$

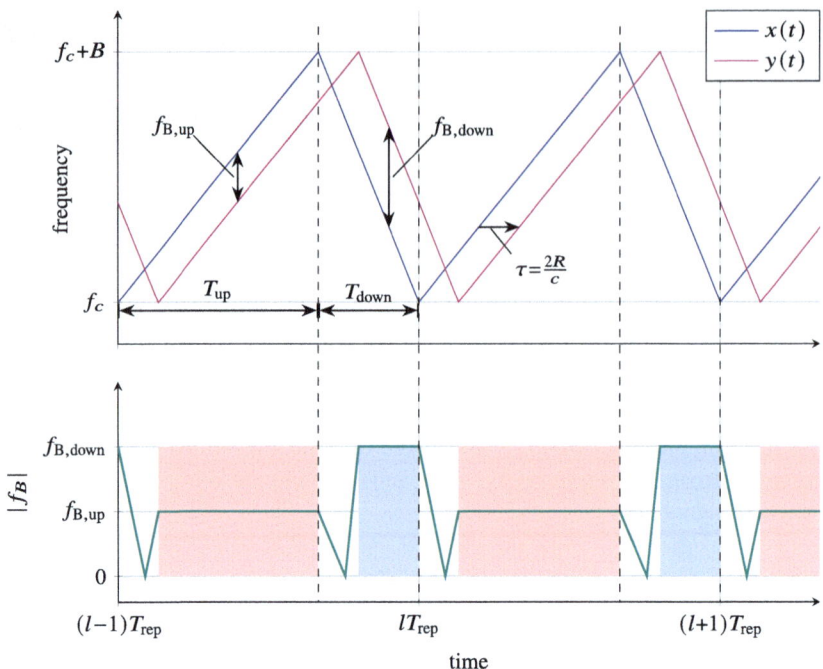

Fig. 4.7 FMCW beat signal for the stationary case ($v \approx 0$ m/s) using a triangular signal shape with rising ramp duration T_{up} and falling ramp duration T_{down} (*upper graph*); The received signal (*red*) is delayed by the time τ compared to the transmitted FMCW signal (*blue*). The resulting beat signal is shown in the *lower graph*. The delay τ causes two beat signal states: $f_{\text{B,up}}$ for the rising and $f_{\text{B,down}}$ for the falling ramp

Figure 4.7 shows a range measurement example using an FMCW signal consisting of a rising ramp of duration T_{up} and a falling ramp of duration T_{down}. The repetition or period duration of the basic signal shape is thus $T_{\text{rep}} = T_{\text{up}} + T_{\text{down}}$. The upper graph in the figure shows the l-th and $(l+1)$-th periods of the transmit and receive ramps. Due to the channel delay of the signal, the receive ramps (*magenta*) are shifted by the duration τ with respect to the transmit ramps (*blue*). In the frequency-time domain, mixing the transmit and receive ramps corresponds to taking the difference between the two frequency responses. This is shown in the lower graph in Fig. 4.7 as the frequency of the beat signal. If the slopes of the rising and falling ramp segments differ, two different beat frequencies are produced, one for the rising ramp

$$|f_{\text{B,up}}| = \frac{2BR}{T_{\text{up}}c} \tag{4.22}$$

and one for the falling ramp

4.1 Analog Modulation Schemes

$$|f_{B,\text{down}}| = \frac{2BR}{T_{\text{down}}c}. \tag{4.23}$$

In the evaluation, only the time intervals of the beat signal can be evaluated in which the transmit and receive ramps have the same slope. This corresponds to the intervals in which the resulting beat frequency is constant, see highlighted areas in Fig. 4.7. This period is referred to as the effective measurement duration T_{meas}. As can be seen in Fig. 4.7, T_{meas} depends on the ramp slope, but also on the delay τ. The steeper the ramp and the longer the signal delay, the shorter T_{meas} and thus the ratio of useful signal duration to total measurement duration. If $T_{\text{meas}} \approx T_{\text{rep}}$ as in short-range applications, almost the entire signal can be evaluated.

4.1.2.2 Range and Velocity Measurement with FMCW Radars

If the range- and velocity-dependent delay time τ in (4.19) is substituted by (4.3), the beat signal phase for moving targets becomes

$$\phi_B(t) \approx 2\pi \left(f_c \frac{2R}{c} + \frac{2vf_c}{c}t \pm \frac{B}{T}\frac{2R}{c}t \pm \underbrace{\frac{B}{T}\frac{2v}{c}t^2}_{\ll} \right). \tag{4.24}$$

The last term is called *range-Doppler coupling* and can usually be neglected, since it is significantly smaller compared to the other terms [4].

The derivative of the phase (4.24) with respect to time yields the beat frequency

$$f_B = \frac{1}{2\pi}\frac{d\phi_B(t)}{dt} = \frac{2vf_c}{c} + \frac{2B}{T}R = f_D \pm \frac{2B}{Tc}R, \tag{4.25}$$

for a rising (+) or falling (−) ramp, which contains both a range-dependent and a velocity-dependent component. It should be noted that the range and velocity information can no longer be unambiguously extracted from the beat signal, since if f_B is known, it is unclear which part is to be attributed to the target range and which to the target velocity. This dilemma is solved by using a triangular signal with identical ramp slopes of opposite signs for the falling and rising ramps, as shown at the top of Fig. 4.8. The received signal is shifted by the range-dependent component $2R/c$ in the x direction and by the velocity-dependent component f_D in the y direction. In this case, $T_{\text{up}} = T_{\text{down}} = T$ and the beat frequencies (4.25) of the rising (+) and falling (−) ramp become

$$f_{B,\text{up}} = f_D + \frac{2B}{Tc}R \tag{4.26}$$

and

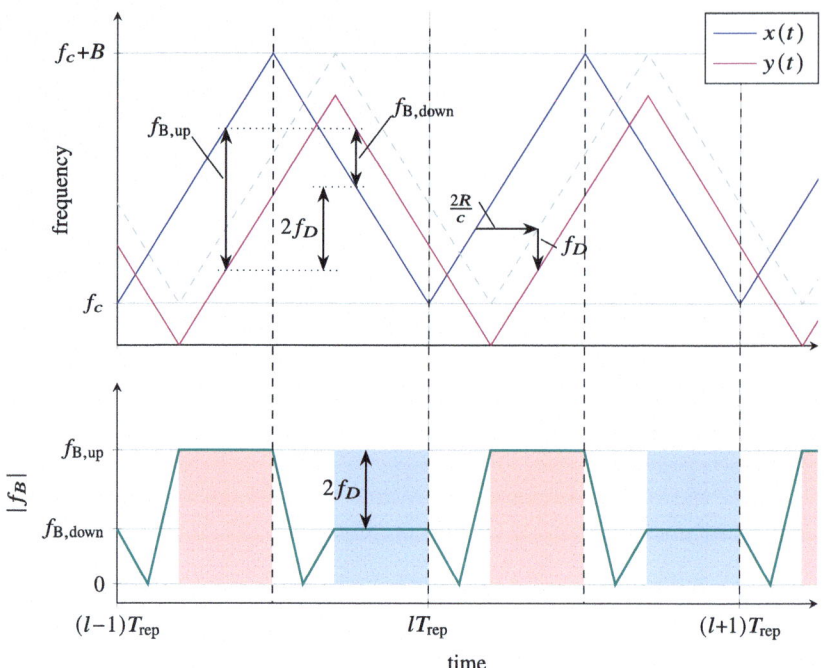

Fig. 4.8 FMCW beat signal with Doppler shift ($v \neq 0$ m/s) using a triangular signal shape with identical rising and falling ramp duration $T = T_{\text{up}} = T_{\text{down}}$ (*upper graph*); Compared to the transmitted signal (*blue*) the receive signal (*red*) is shifted by $2R/c$ in time (*gray dashed line*) due to the target range. Due to the target velocity, the receive signal is additionally shifted by f_D in frequency. In the beat signal (*lower graph*), this leads to two beat signal states: $f_{B,\text{up}}$ for the rising and $f_{B,\text{down}}$ for the falling ramp. Due to the identical slopes, the difference between these frequencies is $2 f_D$

$$f_{B,\text{down}} = f_D - \frac{2B}{Tc} R \,. \tag{4.27}$$

In the case of a stationary target, (4.26) and (4.27) become equal: $f_{B,\text{up}} = f_{B,\text{down}}$; in the case of $v \neq 0$ m/s, they differ. The velocity and range information is thus extracted by calculating the sum and difference of (4.26) and (4.27), which results in

$$f_{B,\text{up}} + f_{B,\text{down}} = 2 f_D \,, \tag{4.28}$$

and

$$f_{B,\text{up}} - f_{B,\text{down}} = 2 \frac{2B}{Tc} R \,, \tag{4.29}$$

respectively. This procedure can be illustrated graphically, as depicted in Fig. 4.9a. If the beat frequencies (4.26) and (4.27) are plotted in an f_D-vs.-f_R- diagram (where

4.1 Analog Modulation Schemes

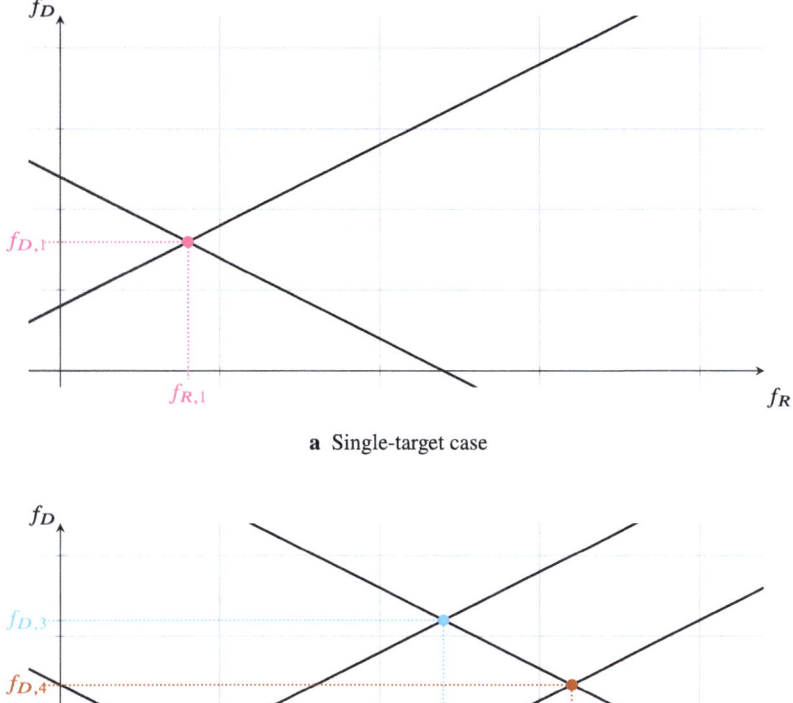

Fig. 4.9 Illustration of the determination of the range and Doppler components **a** of a single target and **b** of two targets measurement using the sum and difference (4.28) and (4.29), respectively, of beat signals of the rising and falling edge. For the single target case, the solution is unique while for two or more targets, the solution becomes ambiguous

$f_R = 2R/c$), an intersection point is obtained that lies at the target parameters $f_{R,1}$ and $f_{D,1}$. However, the procedure is only unambiguous for the single-target case.

In the case of Q target reflections, Q beat frequencies $f_{B,\text{up},i}$, $i = 1, \ldots, Q$, are obtained for the rising ramp and, likewise, Q beat frequencies $f_{B,\text{down},j}$, $j = 1, \ldots, Q$, for the falling ramp. However, the correct mapping between them is not known. With Q different targets, Q^2 possible combinations and thus solutions to (4.28) and (4.29) are obtained, i.e. theoretical target velocities and ranges. This problem is illustrated in Fig. 4.9b for the case of two targets. Four intersections are found between the beat

Table 4.1 Example parametrization for an FMCW radar for measuring the range of a stationary target

Modulation parameters	
f_c	77 GHz
B	2 GHz
T	1 ms
f_s	100 kHz
Performance parameters	
ΔR	0.075 m
R_{ua}	15 m

signals of the rising and falling ramps, with only two of the intersections actually corresponding to targets and two representing ghost targets. One solution for the multi-target case is to use different ramp slopes. Since only the range-dependent frequency component f_R is proportional to the slope B/T, different linear functions are obtained for different ramp slopes. The points of intersection of the linear functions, which correspond to actual targets, do not change as a result, but the incorrect points of intersection shift, which means that correct and incorrect targets can be separated.

4.1.2.3 Example Parametrization for an FMCW Radar

Since FMCW radars are used primarily in stationary scenarios, Table 4.1 shows a typical design of an FMCW radar for a non-moving target. The corresponding performance parameters are derived from the signal processing described in Chap. 5.

4.1.3 Chirp Sequence Radars

With chirp sequence radar, as with FMCW radar, linear frequency ramps are transmitted. These typically have very steep ramp slopes or very short ramp periods (e.g. $T < 50\,\mu s$). In the past, generating fast linear frequency ramps with large bandwidths was a challenge. Today, such signals are realized easily and cheaply by integrated PLLs and fast circuits, so that the chirp sequence radar has become widely used, especially in the automotive industry. In terms of modulation, the chirp sequence radar is a special form of FMCW radar. However, a distinction based solely on the modulation parameters (e.g. the ramp slope) is not clearly defined.

The difference between FMCW and chirp sequence radars however becomes apparent when analyzing the velocity information. Due to the use of very steep ramp slopes, the frequency shift caused by the Doppler effect in the beat signal (4.25) is very small compared to the frequency shift caused by the target range. This makes

4.1 Analog Modulation Schemes

the Doppler shift in (4.25) negligible, so that approximately

$$f_B = \underbrace{f_D}_{\ll} + \frac{2B}{Tc}R \approx \frac{2B}{Tc}R \qquad (4.30)$$

can be assumed for the beat frequency of the chirp sequence radar. As a consequence, it is no longer possible to determine the Doppler shift using a single frequency ramp in this case. Instead, it is detected using a sequence of several frequency ramps, as described in more detail below.

4.1.3.1 Range and Velocity Measurement with Chirp Sequence Radars

To measure the velocity of targets in addition to their range with a chirp sequence radar, a frame of L frequency ramps (4.14), i.e. chirps, is transmitted in succession. The transmit signal is given by

$$x(t) = \sum_{l=0}^{L-1} \cos\underbrace{\left(2\pi\left(f_c t \pm \frac{1}{2}\frac{B}{T}t^2\right)\right)}_{\phi_{tx}(l,t)} \operatorname{rect}\left(\frac{t - lT_{\text{rep}}}{T}\right). \qquad (4.31)$$

The chirp repetition time $T_{\text{rep}} = T + T_{\text{down}}$ is the sum of the duration of the rising chirp $T = T_{\text{up}}$ and the duration of the falling ramp $T_{\text{down}} \ll T$, as shown in Fig. 4.10a. Only the rising ramp is evaluated later, which is why the transmitter can be switched off for the duration of the falling ramp. In addition, a dead time $T_d \ll T$ can be inserted between the signal periods, as shown in Fig. 4.10b. Analogous to (4.16), the baseband of the received signal is given by

$$y(t) = A\sum_{l=0}^{L-1} \cos\underbrace{\left(2\pi\left(f_c(t-\tau) \pm \frac{1}{2}\frac{B}{T}(t-\tau)^2\right)\right)}_{\phi_{rx}(l,t)} \operatorname{rect}\left(\frac{t - lT_{\text{rep}}}{T}\right). \qquad (4.32)$$

As with the FMCW radar, at the receiver, the transmit signal is mixed with the receive signal, resulting in the beat signal

$$y_B(t) = y(t)x^*(t) = A\sum_{l=0}^{L-1} e^{j\phi_B(l,t)} \operatorname{rect}\left(\frac{t - lT_{\text{rep}}}{T}\right). \qquad (4.33)$$

The phase $\phi_B(l,t)$ of the l-th signal period of the beat signal follows from the difference between the phases of the transmitted and received signals:

$$\phi_B(l,t) = \phi_{tx}(l,t) - \phi_{rx}(l,t) \approx 2\pi\left(f_c\tau \pm \frac{B}{T}t\tau\right). \qquad (4.34)$$

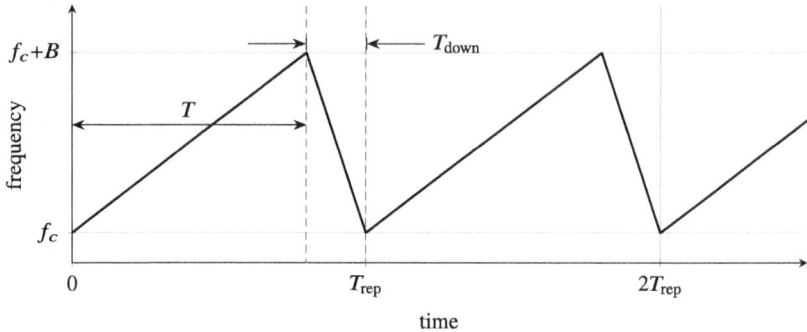

a Signal using continuous chirps

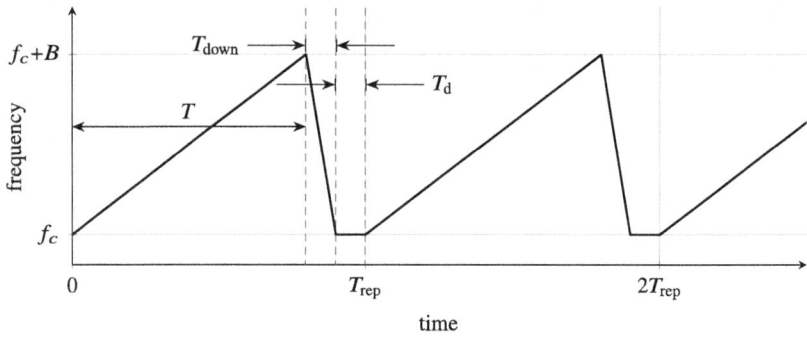

b Signal using chirps with dead time

Fig. 4.10 Chirp sequence frequency ramps with rising chirp duration T and falling chirp duration $T_{\text{down}} \ll T$. **a** Chirp sequence signal without and **b** with additional dead time T_d between the chirps

Again, the phase component proportional to τ^2 can be neglected. Due to the short chirp periods ($T_{\text{rep}} < 100\ \mu s$) and expected (mostly) small relative velocities ($v < 100$ m/s), the change in target range between the $(l-1)$-th and the l-th chirp resulting from the relative velocity between radar and target is significantly smaller than the range resolution ΔR of the radar:

$$R_l - R_{l-1} = v T_{\text{rep}} \ll \Delta R.$$

Consequently, the influence of the velocity on the phase during a single chirp is negligible and the travel time of the signal can be simplified to

$$\tau = \frac{2R}{c} + \frac{2v}{c} l T_{\text{rep}}. \tag{4.35}$$

With this, the phase of the beat signal (4.34) results in

4.1 Analog Modulation Schemes

$$\phi_B(l,t) \approx 2\pi \left(f_c \frac{2R}{c} + \frac{2vf_c}{c}lT_{\text{rep}} \pm \frac{B}{T}\frac{2R}{c}t \pm \underbrace{\frac{B}{T}\frac{2v}{c}lT_{\text{rep}}t}_{\ll} \right). \quad (4.36)$$

The last term can be neglected due to the previously used assumption that $vT_{\text{rep}} \ll \Delta R = c/(2B)$ applies. The derivative of the phase with respect to time t, i.e. the fast time, results in a frequency component

$$f_R = \frac{1}{2\pi}\frac{d\phi_B(l,t)}{dt} = \pm\frac{B}{T}\frac{2R}{c} \quad (4.37)$$

that is solely proportional to the range. Whereas the derivative with respect to slow time l yields the velocity-dependent frequency component

$$f_v = \frac{1}{2\pi}\frac{d\phi_B(l,t)}{dl} = \frac{2vf_c}{c}T_{\text{rep}}. \quad (4.38)$$

f_v depends only on the carrier frequency, but not on the ramp slope, and vice versa, f_R depends only on the ramp slope, but not on the carrier frequency. In summary, the complex beat signal of the chirp sequence radar is given by

$$y_B(t) = A\sum_{l=0}^{L-1} e^{j\phi_B(l,t)} \operatorname{rect}\left(\frac{t-lT_{\text{rep}}}{T}\right) \quad (4.39)$$

$$= A\sum_{l=0}^{L-1} e^{j2\pi\left(f_c\frac{2R}{c}+\frac{2vf_c}{c}lT_{\text{rep}}\pm\frac{B}{T}\frac{2R}{c}t\right)} \operatorname{rect}\left(\frac{t-lT_{\text{rep}}}{T}\right) \quad (4.40)$$

$$= A\sum_{l=0}^{L-1} e^{j2\pi\left(f_c\frac{2R}{c}+f_v l+f_R t\right)} \operatorname{rect}\left(\frac{t-lT_{\text{rep}}}{T}\right). \quad (4.41)$$

Figure 4.11 shows an example of the beat signal and its components. The graph at the top shows the transmit and receive ramps. The receive ramp is shifted by the round-trip time τ in the x direction and by the Doppler frequency in the y direction. The latter is relatively small, which is why it is barely noticeable. The second graph shows the resulting beat frequency. The third graph shows the amplitude of the beat signal, which changes with frequency f_R (4.37). Extracting the initial phase of the ramps across a series of ramps, as shown in the bottom graph, the velocity-dependent component f_v (4.38) can be observed.

4.1.3.2 Example Parametrization for a Chirp Sequence Radar

Table 4.2 shows a typical configuration of a chirp sequence modulation for non-stationary scenarios. Non-stationary means that, in addition to the range, the radial

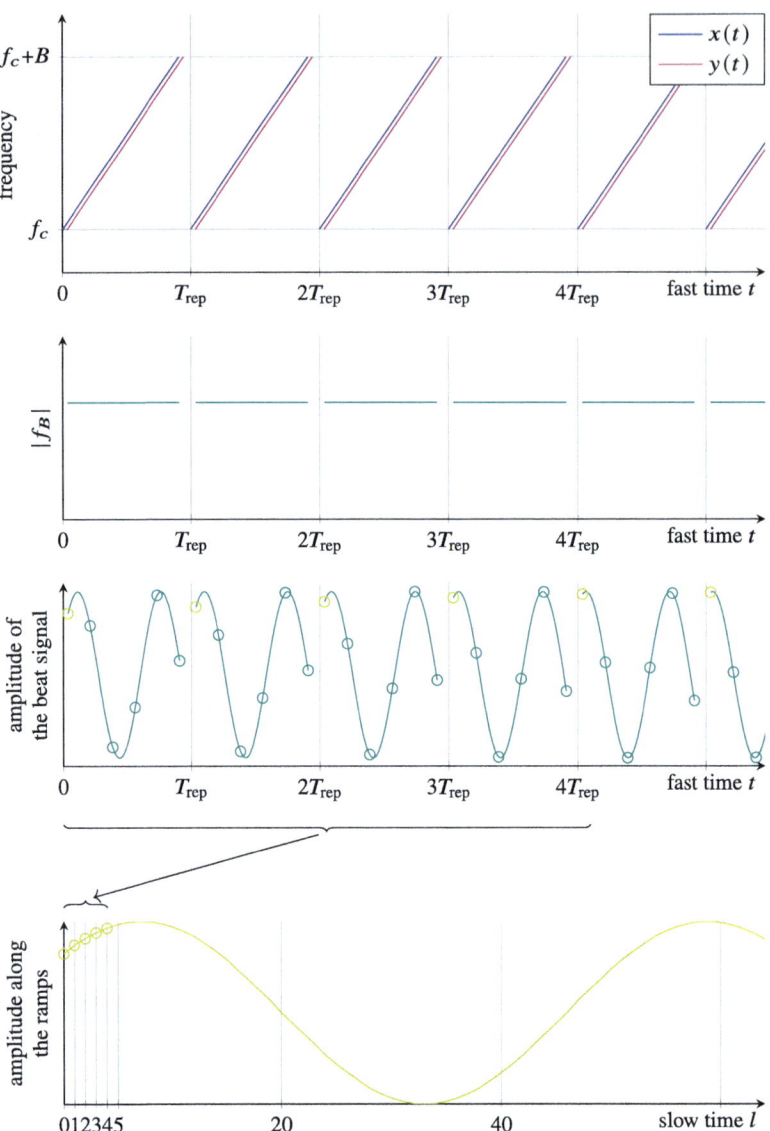

Fig. 4.11 Beat signal of a chirp sequence radar for a moving target ($v \neq 0$ m/s). *First graph:* transmitted (*blue*) and received (*red*) rising chirps; *Second and third graphs:* corresponding beat frequency f_B and amplitude of the beat signal, respectively, for the case of a single moving target. *Forth graph:* amplitude progression in the slow time of the starting amplitudes of the beat signal along several chirps resulting in a wave with frequency f_v (4.38)

4.2 Digital Modulation Schemes

Table 4.2 Example parametrization for a chirp sequence radar for measuring the range and velocity of a moving target

Modulation parameters	
f_c	77 GHz
B	2 GHz
T	30 µs
L	512
f_s	150 MHz
Performance parameters	
ΔR	0.075 m
R_{ua}	337.5 m
Δv	0.126 m/s
v_{ua}	32.46 m/s

velocity of the target relative to the radar is measured. The performance parameters are derived from the signal processing in Sect. 5.3.3.

4.2 Digital Modulation Schemes

Digital modulation techniques offer enormous flexibility in the design of the modulation compared to analog techniques. The modulation can be easily adapted to the measurement scenario, for example to respond flexibly to interference between radar sensors. Sensor architectures that perform signal synthesis and essential signal processing entirely in the digital domain are referred to as digital radars. On the transmitting side, the baseband signal with the bandwidth of the millimeter wave signal is synthesized by digital-to-analog conversion and then mixed into the millimeter wave band. On the receiving side, the millimeter wave band is again mixed into the baseband and converted from analog to digital with the bandwidth of the millimeter wave signal. Therefore, millimeter-wave radar sensors with bandwidths in the gigahertz range require conversion rates of several GS/s, which leads to corresponding data volumes that have to be processed in real time, depending on the application. This requires fast digital circuits and high storage capacity, which generate a correspondingly large amount of power dissipation. Despite these disadvantages, the advantages of digital modulation outweigh those of analog modulation in many applications. Implementations of digital radar sensors in the millimeter wave range with real-time signal processing and bandwidths of several gigahertz on a single field-programmable gate array (FPGA) were already demonstrated around 2020 [5].

In the ongoing discussion, the two digital modulation schemes orthogonal frequency-division multiplexing (OFDM) and phase-modulated continuous-wave (PMCW) are being considered in particular; Both are presented in this section in their basic form. However, it is clear that due to the digital signal generation many variants and combinations of modulation approaches can be implemented very flexibly.

4.2.1 OFDM Radars

The OFDM scheme originally comes from communications engineering. It is a special form of frequency-division multiplexing (FDM). An OFDM signal, that is used to carry out a radar measurement, can therefore be used simultaneously for communications.

An OFDM signal or frame, as used in both communications and radar technology, is characterized by being divided into discrete units in both the time and frequency domains. In the time dimension, an OFDM signal consists of a sequence of L OFDM symbols x_l, where $l = 0, \ldots, L-1$. The OFDM symbols have identical durations T and occupy a frequency spectrum of bandwidth B. In the frequency dimension, the bandwidth B of each OFDM symbol is evenly divided into N so-called subcarriers. The frequency spacing between two adjacent subcarriers is referred to as the subcarrier spacing $\Delta f = B/N$. The subcarriers are orthogonal to each other if the ratio between the symbol duration T and subcarrier spacing Δf is

$$\Delta f = \frac{1}{T}. \tag{4.42}$$

This turns a conventional FDM signal into an OFDM signal. This orthogonality may be best illustrated by the spectrum of an OFDM symbol, as shown in Fig. 4.12. In the spectrum, each subcarrier is described by a sinc function, which has a maximum at the subcarrier frequency $n\Delta f = f_n$, $n = 0, \ldots, N-1$, and zeros at all other subcarrier frequencies $f_{n'}$, $n \neq n'$. The reason for this is that in the time domain, the OFDM symbol is limited to a duration of T, which corresponds to a multiplication of an infinitely long signal with a rectangular function of length T. The corresponding function of the rectangle in the image domain is the sinc function

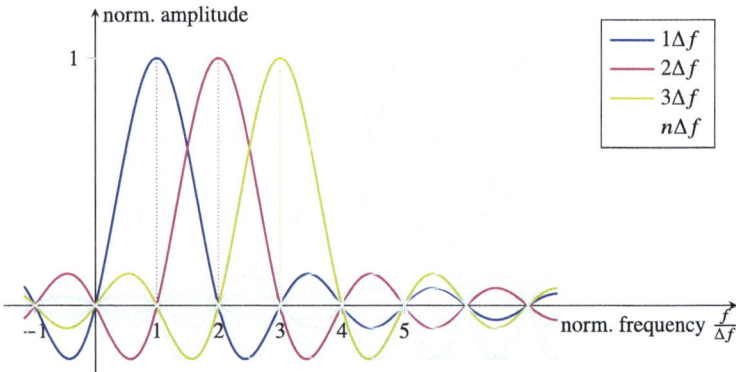

Fig. 4.12 Spectrum of a single OFDM symbol where each subcarrier $n\Delta f$ is described by a sinc function with maximum at its subcarrier frequency and zeros at all other subcarrier frequencies

4.2 Digital Modulation Schemes

$$\text{rect}\left(\frac{t}{T}\right) \overset{\mathcal{F}}{\Longleftrightarrow} |T|\text{sinc}(fT) \qquad (4.43)$$

which has zeros at integer multiples of $1/T$. For a subcarrier defined by the complex oscillation $\exp\{j2\pi f_n t\}$ this yields

$$e^{j2\pi f_n t}\text{rect}\left(\frac{t}{T}\right) \overset{\mathcal{F}}{\Longleftrightarrow} |T|\text{sinc}(fT) * \delta(f-f_n) = |T|\text{sinc}((f-f_n)T). \qquad (4.44)$$

The characteristic orthogonality of the signal is evident in the spectrum in Fig. 4.12 in that the maxima of the subcarriers fall on the zeros of all other subcarriers. For signal transmission on the individual subcarriers, this means that there is no crosstalk between the subcarriers and that they can be unambiguously and easily separated from each other at the receiver. From a mathematical point of view, the subcarriers thus form an orthogonal basis.

Another characteristic of the OFDM signal is that a modulation symbol d_{ln} is transmitted on each subcarrier for the symbol duration T. A variety of discrete modulation alphabets can be used for this. Usually, a variant of phase-shift keying (PSK) is used, for example quadrature phase-shift keying (QPSK). When evaluating radar signals, it is also advantageous to choose a modulation alphabet with a uniform amplitude. Examples of non-uniform amplitude distributions are pure amplitude modulations or amplitude-phase modulations such as quadrature amplitude modulation (QAM). Subsequently, a QPSK modulation with the four phase angles $\phi_{\text{QPSK}} \in \{\pi/4, 3\pi/4, -3\pi/4, -\pi/4\}$ and symbols $\exp\{j\phi_{\text{QPSK}}\}$ is assumed. Figure 4.13 depicts five subcarriers with frequencies $1/T, 2/T, 3/T, 4/T$, and $5/T$; which either are unmodulated ($\phi = 0$, see Fig. 4.13a) or modulated using randomly selected QPSK phases (see Fig. 4.13b).

A single OFDM symbol l of bandwidth B consisting of N subcarriers with spacing Δf is modeled in the time domain by

$$x_l(t) = \sum_{n=0}^{N-1} d_{ln} e^{j2\pi n \Delta f t}, \quad t \in [0, T). \qquad (4.45)$$

However, the signal is initially only available in a discrete-time digital form at the transmitter. Assuming that the sampling rate corresponds to the Nyquist frequency $f_s = N/T$, the discrete-time description of the signal is

$$x_l[k] = \sum_{n=0}^{N-1} d_{ln} e^{j2\pi \frac{nk}{N}}, \quad k = \lfloor t f_s \rfloor, \qquad (4.46)$$

where k describes the sampling points of the l-th OFDM symbol, which is equivalent to the fast time in FMCW radar. Equation (4.46) in turn corresponds to the inverse discrete Fourier transform (IDFT) of the sequence

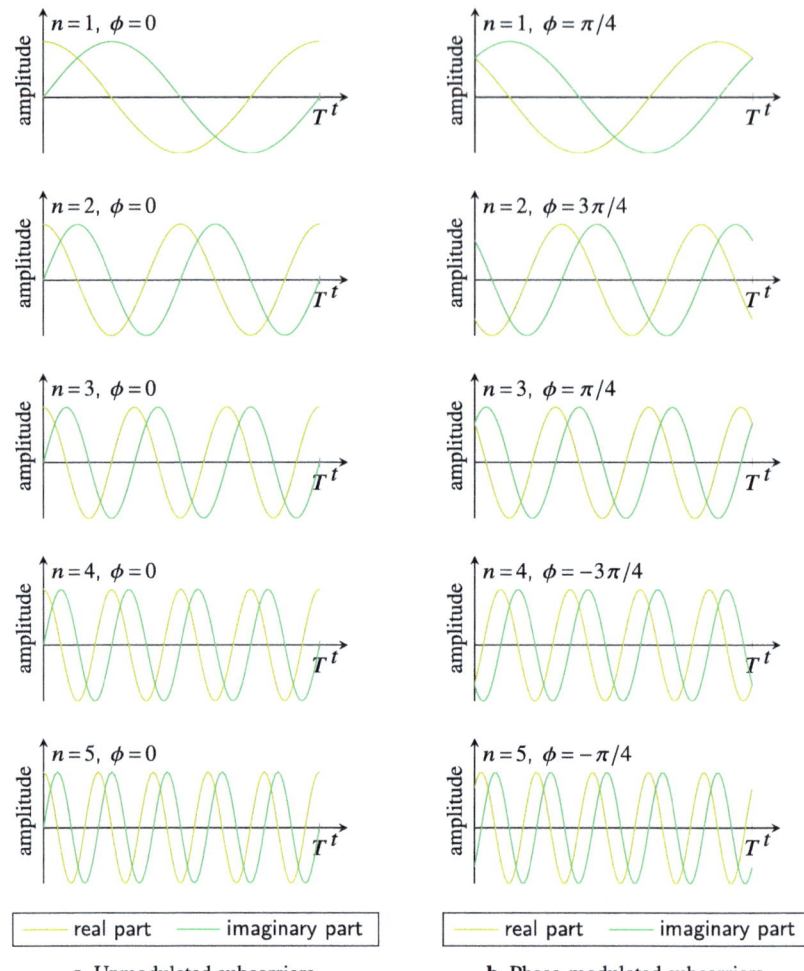

Fig. 4.13 Example of an OFDM modulation for five subcarriers ($n = 1, \ldots, 5$). **a** Real (*yellow*) and complex (*green*) component of the subcarriers without modulation and **b** with phase modulation using the modulation phase ϕ. Compared to the unmodulated waves, the subcarriers initial phase is shifted by ϕ

$$\mathbf{d}_l = [d_{l0}, d_{l1}, \ldots, d_{l(N-1)}]$$

of the modulation symbols of the l-th OFDM symbol:

$$x_l[k] = \text{IDFT}\{\mathbf{d}_l\}. \tag{4.47}$$

This allows the individual OFDM symbols to be efficiently synthesized digitally from (randomly chosen) modulation symbols. If the same signal (4.45) is to be

4.2 Digital Modulation Schemes

generated directly in analog form, a frequency synthesizer with filtering, a phase shifter and a mixer would be required for each of the N subcarriers. Since the number of subcarriers required in radar operation is very large (>1000), a purely analog signal generation is not feasible.

In radar operation, an OFDM frame composed of a continuous sequence of L OFDM symbols (4.46) is transmitted. The choice of the transmit symbols of the individual subcarriers and OFDM symbols is arbitrary and usually random. Due to different delays of the received signals caused by different path lengths of the reflections as well as signal distortions caused by non-ideal filters, intersymbol interference (ISI) occurs in the receiver. This effect describes the unwanted smearing and crosstalk of consecutive symbols in the evaluation. To prevent ISI, a so-called *cyclic prefix* is added before each OFDM symbol in the time domain. The cyclic prefix is a copy of duration $0 \leq T_{cp} \leq T$ of the end of the respective OFDM symbol. It therefore increases the total symbol duration to $T_{OFDM} = T + T_{cp}$. Furthermore, the transmit signal becomes quasi-cyclic and the receive signal becomes a cyclic convolution of the transmit signal with the channel impulse response. For practical reasons, the length of the prefix is chosen as a multiple of the sampling interval $1/f_s$. In the discrete-time case, the prefix is therefore a copy of the last $N_{cp} = T_{cp}f_s$ samples $(x_l[N - N_{cp} - 1], \ldots, x_l[N - 1])$ of (4.46). For better readability of the signal, the prefix is neglected in the mathematical description below.

The digitally generated sequence of L OFDM symbols of bandwidth B is fed to a digital to analog converter (DAC) and converted at a sampling frequency of $f_s \geq B$. At the radar front end, the baseband signal is simply amplified, mixed to the millimeter-wave (RF) carrier frequency f_c and transmitted. The transmitted OFDM symbol is described by

$$x_l^{RF}(t) = \sum_{n=0}^{N-1} d_{ln} e^{j2\pi(n\Delta f + f_c)t}, \quad t \in [0, T). \tag{4.48}$$

The signal detected at the receiver is shifted by the delay τ (4.3). It is first mixed back to the baseband with the same monotone carrier signal $\exp\{-j2\pi f_c t\}$:

$$\begin{aligned} y_l(t) &= x_l^{RF}(t - \tau) e^{-j2\pi f_c t} \\ &= \sum_{n=0}^{N-1} d_{ln} e^{j2\pi n\Delta f t} e^{-j2\pi(n\Delta f + f_c)\tau(t)}, \quad t \in [0, T). \end{aligned} \tag{4.49}$$

The signal is digitized at the analog to digital converter (ADC) and then split into symbols of length T_{OFDM}. The first N_{cp} samples of each symbol are usually discarded, which removes the cyclic prefixes. Figure 4.14 visualizes the shift between the transmitted and received signal as well as the signal portion used for radar evaluation. With $t = kT/N + lT_{OFDM}$, this gives the discrete-time baseband signal

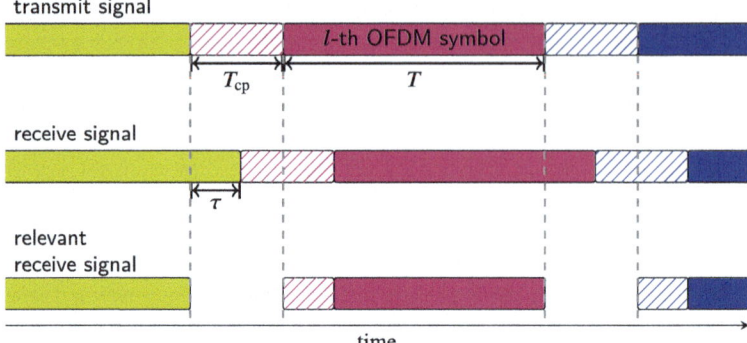

Fig. 4.14 Illustration of a transmit and receive sequence for a series of OFDM symbols (*yellow, red, blue*) where only the middle (*l*-th) symbol is shown entirely in the graph. The receive frame (*middle row*) is shifted by τ compared to the transmit frame (*top row*). At the receiver, the part of the receive signal where the cyclic prefix was transmitted in the transmit signal is removed (*bottom row*). If the delay τ is not longer than the cyclic prefix duration T_{cp}, each of the resulting signal chunks contains only the parts of a single transmitted OFDM symbol

$$y_l[k] = \sum_{n=0}^{N-1} d_{ln} e^{j2\pi \frac{kn}{N}} e^{j\phi_{ln}^{rx}[k]} \qquad (4.50)$$

where the receive phase $\phi_{ln}^{rx}[k]$ of the l-th symbol and n-th subcarrier is

$$\phi_{ln}^{rx}[k] = 2\pi \Big(-f_c \frac{2R}{c} - n\Delta f \frac{2R}{c} + f_D l T_{OFDM} \\ + \underbrace{f_D \frac{kT}{N}}_{\ll} - \underbrace{n\Delta f \frac{2v}{c} \frac{kT}{N}}_{\ll} - \underbrace{n\Delta f \frac{2v}{c} l T_{OFDM}}_{\ll} \Big). \qquad (4.51)$$

The first phase term in (4.51) is independent of the slow time l and the subcarrier n and depends only on the range R. The second and third terms are used for range and velocity evaluation, respectively. The last three terms can be neglected in most cases [6], making the phase ϕ_{ln}^{rx} in (4.50) independent of the fast time variable k. Therefore, the receive phase often is simplified to

$$\Rightarrow \phi_{ln}^{rx}[k] \approx \phi_{ln}^{rx} = 2\pi \Big(-f_c \frac{2R}{c} - n\Delta f \frac{2R}{c} + f_D l T_{OFDM} \Big). \qquad (4.52)$$

In the next step, a DFT of the discrete-time baseband signal (4.50) is evaluated along the fast time (k). This gives the spectra of the individual OFDM symbols

$$Y[l, n] = \text{DFT}\{\mathbf{y}_l\} = d_{ln} e^{j\phi_{ln}^{rx}} \qquad (4.53)$$

4.2 Digital Modulation Schemes

where $\mathbf{y}_l = (y_l[0], \ldots, y_l[N-1])$. The length of the DFT is therefore T, hence the frequency spacing of the spectral coefficients of the discrete spectrum is again $\Delta f = 1/T$. Thus, the DFT can be used to de-multiplex the signal back into its subcarriers. The elements $Y[l,n]$ (4.53) of the received signal in the time-frequency domain can be represented in a matrix $\mathbf{Y}$ of dimension $L \times N$. To be able to evaluate the range- and velocity-dependent receive phase ϕ_{ln}^{rx}, the modulation phases d_{ln} must be removed. Since these are known in radar operation, this can be achieved by a complex division of (4.53) and the symbol d_{ln}. Equivalently, if $|d_{ln}|=1\ \forall l, n$ applies, e.g. when using QPSK modulation, (4.53) may just as well be multiplied with the complex-conjugate symbol:

$$D[l,n] = \frac{Y[l,n]}{d_{ln}} \underbrace{\left(= Y[l,n] d_{ln}^* \right)}_{\text{if } |d_{ln}|=1} = e^{j\phi_{ln}^{rx}}. \tag{4.54}$$

4.2.1.1 Range and Velocity Measurement with OFDM Radars

The channel information $D[l,n]$ (4.54) of the l-th OFDM-symbol and n-th subcarrier extracted at the receiver can be represented as $L \times N$ matrix

$$\mathbf{D} = \begin{bmatrix} D[0,0] & \cdots & D[0, N-1] \\ \vdots & \ddots & \vdots \\ D[L-1,0] & \cdots & D[L-1, N-1] \end{bmatrix}. \tag{4.55}$$

A closer look at the receive phase (4.52) shows that the range-dependent component depends only on the subcarrier variable n. Thus, there is a range-dependent behavior along the rows of $\mathbf{D}$. The derivative of the receive phase (4.52) with respect to n yields the range-proportional frequency component

$$f_R = \frac{1}{2\pi} \frac{d\phi_{ln}^{rx}}{dn} = -\Delta f \frac{2R}{c}, \tag{4.56}$$

which can be observed along the rows of $\mathbf{D}$. Conversely, the velocity-dependent component of the phase is only dependent on the slow time variable l, resulting in a velocity-proportional behavior only along the columns of $\mathbf{D}$. The derivative with respect to the slow time l forms the velocity-dependent frequency component

$$f_v = \frac{1}{2\pi} \frac{d\phi_{ln}^{rx}}{dl} = \frac{2v f_c}{c} T_{\text{OFDM}}, \tag{4.57}$$

which can be observed along the columns of $\mathbf{D}$. Due to the linear and orthogonal dependency of both components, they can be evaluated separately.

Table 4.3 Example parametrization for an OFDM radar for measuring the range and velocity of a moving target

Modulation parameters	
f_c	77 GHz
B	1.024 GHz
Δf	500 kHz
N	2048
T	2 µs
L	3072
T_{cp}	0.5 µs
Performance parameters	
ΔR	0.15 m
R_{ua}	300 m
Δv	0.26 m/s
v_{ua}	394.4 m/s

4.2.1.2 Example Parametrization for an OFDM Radar

Table 4.3 shows an example of the modulation parameters of a radar with OFDM modulation. In the non-stationary scenario, the radial velocity relative to the target is to be measured in addition to the range. In this case, the sampling rate f_s must be at least twice as high as the bandwidth and leads to enormous digital data rates. The performance parameters are derived from the signal processing in Sect. 5.3.4.

4.2.2 PMCW Radars

The PMCW radar uses a phase-modulated CW carrier signal that is modulated with a code sequence C. On the transmitter side, a code generator provides a code sequence of fixed length, which is simply modulated onto the millimeter-wave carrier frequency. A typical structure of a PMCW radar is shown in Fig. 4.15. Thus, the transmit signal of the l-th sequence is given by

$$x_l(t) = x^{\text{PN}}(t) \cdot x^{\text{CW}}(t) = x^{\text{PN}}(t) \cdot A_{\text{tx}} \cos\left(2\pi f_c t\right). \tag{4.58}$$

The code sequence C is of duration T and consists of K symbols; This results in the symbol duration $T_c = T/K$. The bandwidth of the transmitted signal is determined by the symbol rate as

$$B = \frac{1}{T_c} = \frac{K}{T}. \tag{4.59}$$

4.2 Digital Modulation Schemes

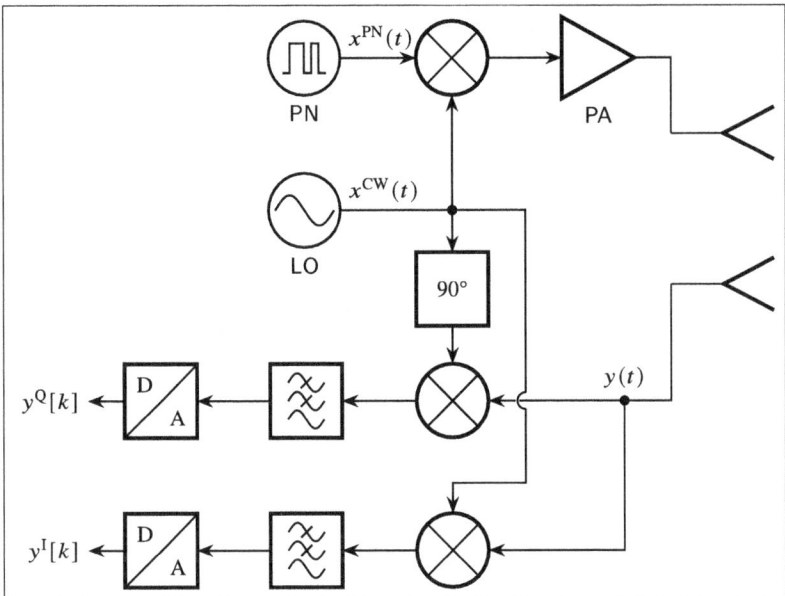

Fig. 4.15 Block diagram of a PMCW radar. The block diagram of a digital radar is very different from that of an analog radar. The baseband signals are only mixed into the millimeter wave range or are mixed from the millimeter wave range into the baseband. On the receive side, an IQ mixer is used to evaluate the complex-valued signal

An essential requirement for the code sequence is a good autocorrelation with a narrow, high peak at lag 0 and otherwise low amplitude close to 0. In particular, in MIMO operation or when operating several radar sensors in parallel, the cross-correlation of different sequences plays a crucial role. To avoid interference or false detections, the cross-correlation of different sequences should be as small as possible. Suitable code sequence types are, for example binary *maximum length sequences*, *Gold sequences* or *Kasami sequences*. These sequences exist only for lengths of $2^b - 1$, $b \in \mathbb{N}$, where there might not be a sequence for every order b. Figure 4.16 shows a binary Gold sequence of length 63 ($b = 6$) with binary value set, i.e. alphabet $\{-1, 1\}$. The code types differ significantly in their auto- and cross-correlation properties, as can be seen in the Figs. 4.17 and 4.18 using the example of sequences of length 63. Maximum length sequences achieve an optimal autocorrelation, but exhibit slightly poorer cross-correlation properties compared to Gold and Kasami sequences. In contrast, Gold and Kasami sequences have significantly worse autocorrelation properties, which plays an important role in range evaluation, as shown in Sect. 5.3.5.

In radar operation, the selected sequence is repeated L times within a frame, whereby no dead times are needed between repetitions to account for settling processes or channel delays.

The signal of the l-th sequence detected at the receiver

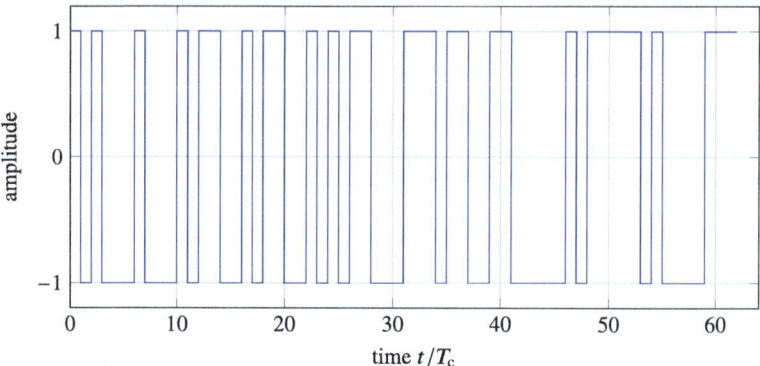

Fig. 4.16 Binary Gold sequence of length 63 with chip rate, i.e. symbol duration T_c and alphabet $\{-1, 1\}$

$$y_l(t) = x_l(t - \tau) = x^{\text{PN}}(t - \tau) \cdot x^{\text{CW}}(t - \tau) \tag{4.60}$$

is delayed by the round-trip time τ (4.3). As shown in Fig. 4.15, the signal is then fed to an IQ mixer, filtered through an anti-aliasing lowpass filter and sampled. This results in the discrete-time in-phase and quadrature components

$$y_l^{\text{I}}[k] = A_{\text{rx}} \cos\left(2\pi f_c k T_s\right) \cdot x^{\text{PN}}[kT_s - \tau] \cdot \cos\left(2\pi f_c (kT_s - \tau)\right) \tag{4.61}$$

and

$$y_l^{\text{Q}}[k] = A_{\text{rx}} \sin\left(2\pi f_c k T_s\right) \cdot x^{\text{PN}}[kT_s - \tau] \cdot \cos\left(2\pi f_c (kT_s - \tau)\right), \tag{4.62}$$

respectively. Together, these yield the complex baseband signal

$$y_l[k] = y_l^{\text{I}}[k] + \mathrm{j} y_l^{\text{Q}}[k] = A_{\text{rx}} x^{\text{PN}}[kT_s - \tau] \cdot \mathrm{e}^{\mathrm{j}2\pi f_c \tau} \tag{4.63}$$

$$= A_{\text{rx}} x^{\text{PN}}[kT_s - \tau] \cdot \mathrm{e}^{\mathrm{j}2\pi (f_c \frac{2R}{c} - f_D k T_s)} \tag{4.64}$$

taking into account the lowpass filter. Similarly to the chirp sequence radar, with a PMCW radar, the phase change due to the Doppler frequency along the samples of a single code sequence is negligible. For this reason, with PMCW radar, this phase is also observed along L code sequences of a frame. The code rate at which the codes are transmitted is $1/T = 1/(T_c K)$. This yields the baseband signal of the l-th sequence

$$y_l[k] = A_{\text{rx}} x^{\text{PN}}[kT_s - \tau] \cdot \mathrm{e}^{\mathrm{j}2\pi (f_c \frac{2R}{c} - f_D l T)}. \tag{4.65}$$

4.2.2.1 Range and Velocity Measurement with PMCW Radars

Comparing the receive signals of the PMCW radar (4.65) with those of a CW radar (4.65) shows that they only differ by the additional modulation term $x^{\text{PN}}[kT_s - \tau]$ of PMCW. As with CW radar, the last term in (4.65) only allows the velocity to be

4.2 Digital Modulation Schemes

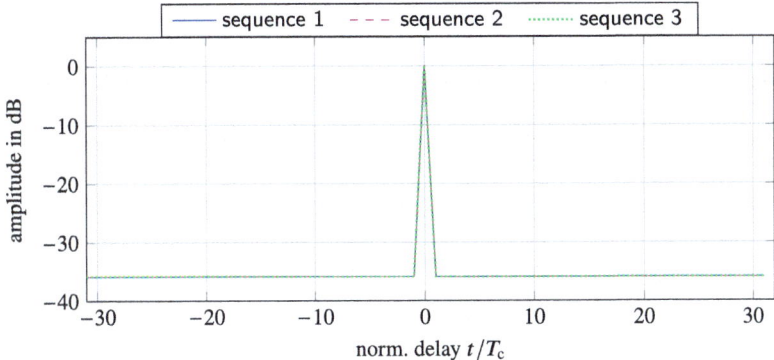

a Maximum length sequence

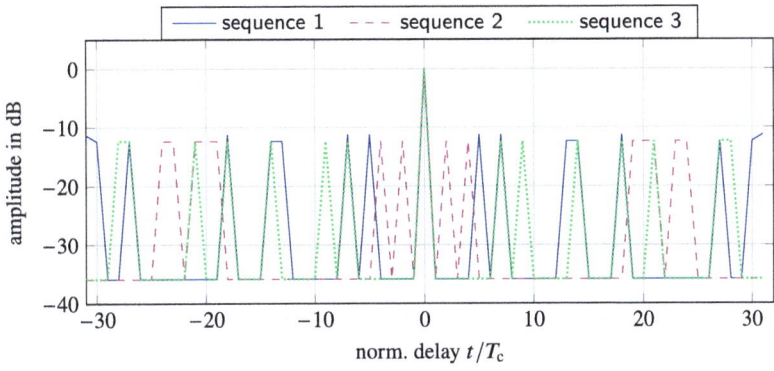

b Gold sequence

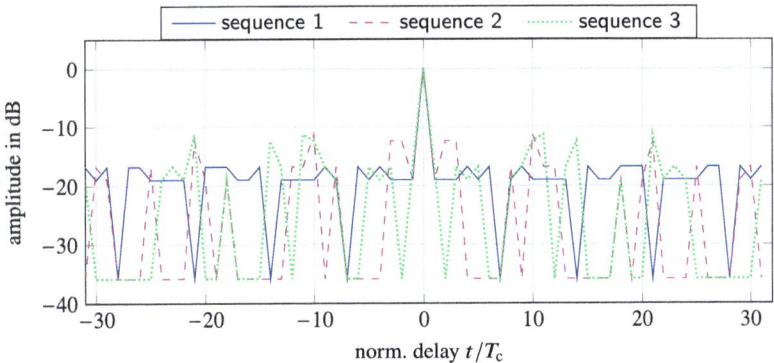

c Kasami sequence

Fig. 4.17 Autocorrelations of each three (*blue, solid; magenta, dashed; green, dotted*) different **a** maximum length, **b** Gold, and **c** Kasami sequences of length 63. Maximum length sequences achieve an optimal autocorrelation with constant side lobes (approx. −36 dB) while both Gold and Kasami sequence achieve only a level of approx. −10 dB for this sequence length

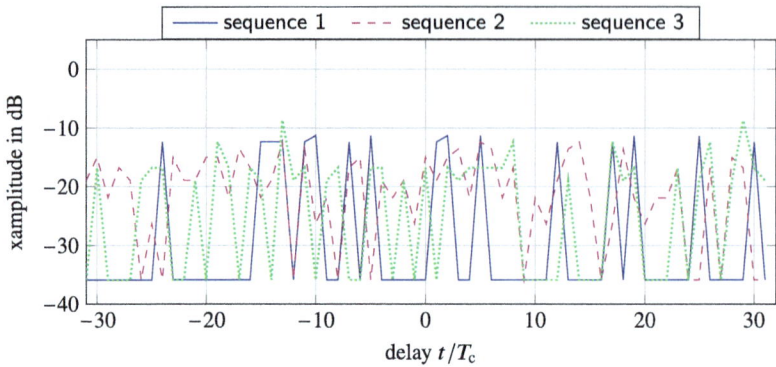

a Maximum length sequence

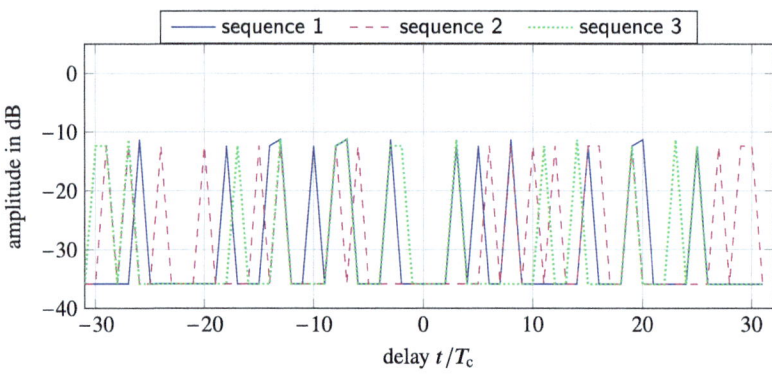

b Gold sequence

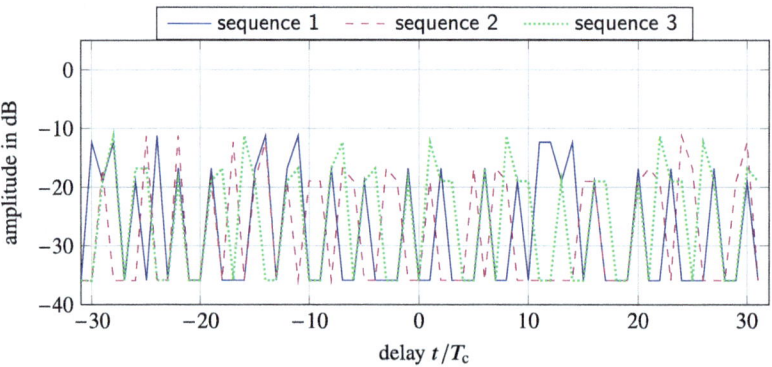

c Kasami sequence

Fig. 4.18 Cross-correlation of each three (*blue, solid; magenta, dashed; green, dotted*) different **a** maximum length, **b** Gold, and **c** Kasami sequence of length 63. All sequences only achieve a cross-correlation, i.e. isolation level of approx. -10 dB

Table 4.4 Example parametrization for a PMCW radar for measuring range and velocity of a moving target

Modulation parameters	
f_c	77 GHz
B	1.024 GHz
T	2 μs
K	2048
L	2024
Performance parameters	
ΔR	0.15 m
R_{ua}	300 m
Δv	0.475 m/s
v_{ua}	475 m/s

determined; but only to a limited extent the determination of the range. In PMCW radar, the range is determined by the modulation term. Due to the propagation time, a range-dependent shift of the modulation sequence is observed at the receiver. If only the time period of a single sequence is considered, the following approximation applies: $\tau \approx 2R/c$. By comparing the received code sequence with the transmitted one by means of correlation, this shift and thus the target range can be determined. The velocity is determined—analogous to a chirp sequence radar—by evaluating the complex phases of the received signal over L sequences of the frame.

4.2.2.2 Example Parametrization for a PMCW Radar

An exemplary design of a PMCW radar for range and velocity measurement is shown in Table 4.4, where the performance parameters are derived from the signal processing in Sect. 5.3.5. As with the OFDM radar, enormous sampling rates may occur.

References

1. J. Hasch, E. Topak, R. Schnabel, T. Zwick, R. Weigel, C. Waldschmidt, Millimeter-wave technology for automotive radar sensors in the 77 GHz frequency band. IEEE Trans. Microw. Theory Tech. **60**(3), 845–860 (2012). https://doi.org/10.1109/tmtt.2011.2178427
2. M. Hitzler, S. Saulig, L. Boehm et al., Ultracompact 160-GHz FMCW radar MMIC with fully integrated offset synthesizer. IEEE Trans. Microw. Theory Techn. **65**(5), 1682–1691 (2017). https://doi.org/10.1109/tmtt.2017.2653111
3. W. Deng, H. Jia, B. Chi, Silicon-based FMCW signal generators: a review. J. Semicond. **41**(11), 111 401 (2020). https://doi.org/10.1088/1674-4926/41/11/111401

4. V. Winkler, Range doppler detection for automotive FMCW Radars, in *European Radar Conference*. (IEEE, 2007). https://doi.org/10.1109/eurad.2007.4404963
5. B. Schweizer, A. Grathwohl, G. Rossi et al., The fairy tale of simple all-digital radars: how to deal with 100 Gbit/s of a digital millimeter-wave MIMO radar on an FPGA [Application Notes]. IEEE Microw. Mag. **22**(7), 66–76 (2021). https://doi.org/10.1109/MMM.2021.3069602
6. C. Knill, Novel MIMO OFDM waveform designs and high-performance signal processing methods for digital radars. Ph.D. Dissertation (2022). https://doi.org/10.18725/OPARU-42838

Open Access This chapter is licensed under the terms of the Creative Commons Attribution 4.0 International License (http://creativecommons.org/licenses/by/4.0/), which permits use, sharing, adaptation, distribution and reproduction in any medium or format, as long as you give appropriate credit to the original author(s) and the source, provide a link to the Creative Commons license and indicate if changes were made.

The images or other third party material in this chapter are included in the chapter's Creative Commons license, unless indicated otherwise in a credit line to the material. If material is not included in the chapter's Creative Commons license and your intended use is not permitted by statutory regulation or exceeds the permitted use, you will need to obtain permission directly from the copyright holder.

Chapter 5
Signal Processing Principles

Building on the introduction to the various radar types and modulation schemes in Chap. 4, this chapter focuses on their signal processing. Here, the range and velocity information is estimated from the signals received by the radar sensors and assigned to the targets. For this purpose, different signal processing chains are used, each involving similar signal processing steps. The usual steps that most processing chains contain are:

1. Windowing
2. Range estimation → DFT along a signal period
3. Velocity estimation → DFT along several signal periods of a frame
4. Angle estimation → evaluation along several channels
5. Target estimation using a CFAR detector

The order of the Fourier transforms, i.e. steps 2 and 3, is permutable due to linearity. Generally, the individual processing steps can be carried out in varying orders, depending on the radar type and application. Two commonly used processing chains are given by the flow charts of Fig. 5.1.

Often, a 2D Fourier transform along the frame's fast time and slow time is first used to generate the so-called range-velocity matrix (R-v matrix), see the flow chart in Fig. 5.1a. If the radar has sufficient range or velocity resolution, range and velocity information of potential targets can be extracted in the R-v matrix using a threshold filtering via a CFAR detection algorithm. Subsequently, an angle estimation is carried out for these targets using the individual radar channels or receiving antennas, as described in Chap. 7.

If the radar has a good separability in velocity but not in range (or vice versa), instead of evaluating the R-v matrix, the dimension with the best separability is first evaluated using a Fourier transform and then an angle estimation is performed, as shown in the flow chart in Fig. 5.1b. Then, a target detection is carried out for the resulting velocity-angle or range-angle image, followed by the estimation of the remaining parameter (range or velocity) for all detected targets.

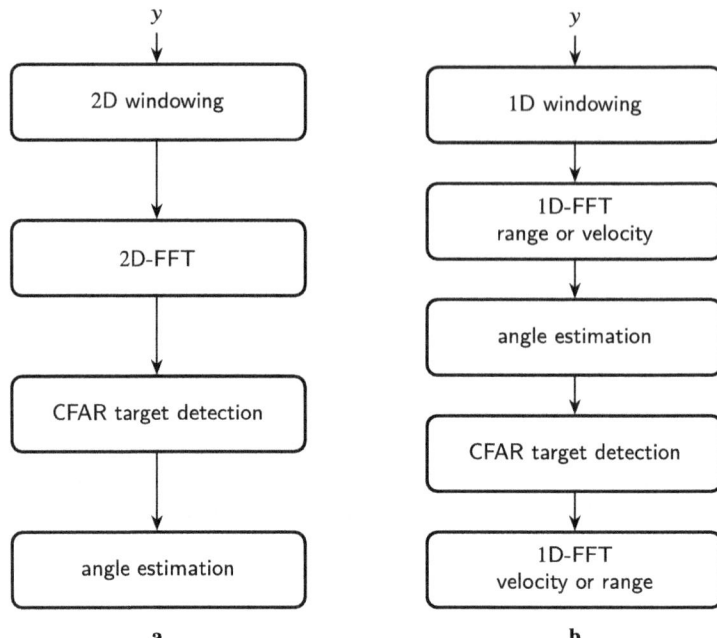

Fig. 5.1 Two common radar signal processing chains for a receive signal y. **a** Processing chain if target separation is performed based on the extracted range and velocity information. In this case, first the range and velocity of the targets is evaluated and the targets are detected based on the R-v matrix. Afterwards, angle estimation is performed for these detected targets. **b** Processing chain if target separation is based on either range or velocity and direction of arrival information. In this case, first either range or velocity evaluation is performed. Afterwards, the angle dimension is evaluation. Based on the angle and range or velocity information, the targets are detected. Finally, the remaining velocity or range of the detected targets is estimated

In this chapter, initially some common terms and key quantities that play an important role in radar signal processing, such as the aforementioned R-v matrix or the CFAR detector, are explained in Sect. 5.1. The following sections explain the individual processing steps in detail: First, in Sect. 5.2, various window types and their significance in signal processing are introduced. After that, the estimation of the range and velocity information for the analog and digital modulation methods introduced in Chap. 4 is presented in Sect. 5.3; The evaluation of the angle information has some special features and is therefore treated separately in Chap. 7. Finally, in Sect. 5.4, target detection using a CFAR detector is presented.

5.1 Key Quantities and Terms

First, some terms that are frequently used in the context of radar signal processing shall be briefly introduced:

Matched Filter A matched filter (MF) maximizes the SNR at the filter output and in this respect represents an optimal linear filter for a defined input signal with additive noise.

R-v Matrix In many applications, the R-v matrix is an essential element in the signal processing chain. It is also referred to as radar image or range-Doppler matrix. Regardless of the radar type (with the exception of the range evaluation for PMCW), it is obtained from the digitized and pre-processed received data by two Fourier transforms. For most radar modulations, the Fourier transform corresponds approximately or exactly to the MF for the respective parameter estimation. The R-v matrix is usually displayed in a grid-like fashion as a matrix grid with the dimensions velocity v and range R and discrete resolution cells, or in short, R-v cells, of size $\Delta R \times \Delta v$. Figure 5.2 shows an R-v matrix of a radar measurement of a passing car. Due to the Fourier transform, the received target power in the signal is concentrated in the R-v cells, which correspond to the measured target ranges and velocities. The noise power, on the other hand, is distributed across all cells. If the target power in a cell is higher than the noise power after integration by the Fourier transform, a peak emerges from the 2D

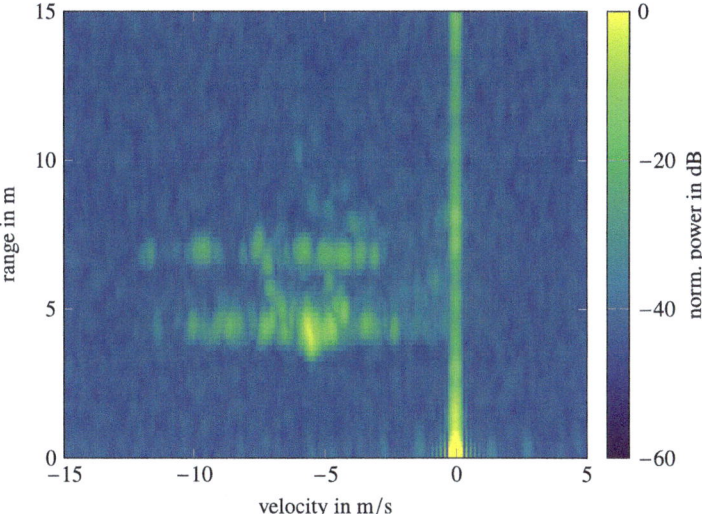

Fig. 5.2 Example of an R-v matrix: Measurement of an approaching car measured by an OFDM radar. The high amplitudes (*green and yellow*) between approx. −3 to −12 m/s and approx. 4–8 m are caused by the car while the high amplitudes for $v \approx 0$ m/s are mostly caused by stationary objects in the surroundings since the observing radar itself is not moving

noise floor at this position. This allows potential targets in the R-v matrix in Fig. 5.2 to be distinguished from the noise (*blue*) by their higher power (*yellow*).

Windowing The finite-length Fourier transform results in high side lobes in the generated spectrum, which can affect target detection and potentially cause false or missed detections. These side lobes can be reduced by windowing, i.e. weighting the processed data before the Fourier transform, thus improving target detection.

Side Lobes Side lobes are understood to be all periodic local maxima that occur in the spectrum in addition to the main lobe. Their position, height and shape depend on the window function used. The highest side lobe usually is the closest to the main lobe or maximum; It determines the side lobe level in decibels or the peak-to-side lobe ratio (PSLR) of the spectrum, as marked Fig. 5.3. For reliable and robust detection of two or more targets in the radar image, the level difference between the strongest and any other target must not be greater than the side lobe level of the strongest target. Otherwise, the weaker target(s) cannot be distinguished from the side lobe of the strongest one.

Integration Gain In a first approximation, radar signals are subject to additive noise, whereby the noise is assumed to be white. The received signal is therefore modeled by

$$x[k] = \tilde{x}[k] + n_0[k] \tag{5.1}$$

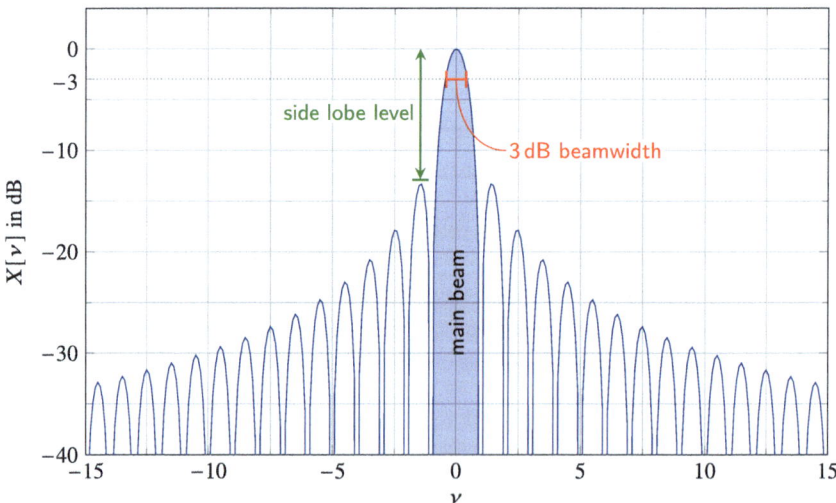

Fig. 5.3 Illustration of the main lobe (beam) (*blue filling*) width (*red*) and side lobe level (*green*) using the example of the sinc-shaped spectrum of the rectangular window

5.1 Key Quantities and Terms

where $\tilde{x}[k]$ is the desired receive signal and $n_0[k]$ describes the additive white noise. White noise is invariant to time and frequency and has zero mean with variance σ^2. Through the Fourier transform, the signal components of a target are constructively superimposed, giving a peak in the spectrum at the respective range and velocity of the target. For a cell v that contains a target, applies

$$\tilde{X}[v] = \mathcal{F}\{\tilde{x}[k]\} = \sum_{k=0}^{K-1} \tilde{x}[k] e^{-j2\pi \frac{kv}{K}} = K E_0 \qquad (5.2)$$

where the expected value of the signal $\tilde{x}$ is

$$E_0 = \mathbb{E}\{|\tilde{x}[k]|\}.$$

The factor K in (5.1) causes the integration gain I of the Fourier transform. On the other hand, the noise power is distributed uniformly across all cells (the Fourier transform of a Gaussian distribution again gives a Gaussian distribution with variance $1/\sigma^2$). Thus, there is no integration gain for the noise component. The SNR in the image domain, i.e. after the Fourier transform, is therefore

$$\text{SNR}_{\text{out}} = I \cdot \text{SNR}_{\text{in}}. \qquad (5.3)$$

Furthermore, each evaluated signal dimension provides an additional integration gain.

CFAR Using a constant false alarm rate (CFAR) detection algorithm, the radar system independently determines which targets are present in the radar measurement. This target detection can take place at various stages in the signal processing chain. It often follows the step of calculating the R-v matrix, in which the targets are detected. An angle estimation is then carried out for each detected target. The prior detection significantly reduces the effort required for the angle estimation. The basic principle of a CFAR detector corresponds to a threshold comparison in which each individual cell or pixel in the radar image is compared with the estimated, scaled local average signal level (background). If the value of the cell exceeds the threshold, it is highly likely to contain a target and is classified as a target cell. The thresholding of a CFAR algorithm is designed to achieve a certain false alarm probability (detection of a target when no target is actually present).

Angle Estimation Through angle estimation, the spatial direction of arrival of a reflection is evaluated or determined. Depending on the processing chain, the angle estimation is carried out at different stages of the chain, see Fig. 5.1. If the R-v matrix is determined first, followed by target detection using a CFAR algorithm, then an angle estimation is carried out only for R-v cells that contain a target. Various algorithms can be used for angle estimation, a selection of which are presented in Chap. 7.

Target List After analyzing the received signal in terms of range and velocity and subsequent target detection using CFAR detection, a list of potential targets can be created. The list contains the range and velocity information of each target that has been detected by the CFAR detector and, if applicable, a subsequent peak detection in the R-v matrix. If an angle estimate has been carried out as well, the list also contains angle information for each target. The target list is then passed on to the next processing stage and can be used, for example, to classify targets (car, pedestrian, cyclist, tree, etc.) or to track targets.

5.2 Windowing

Windowing of the radar signal is an essential step in almost every radar signal processing chain. Applied before an FT, it serves to reduce side lobes caused by the Fourier transform.

Although the FT in radar signal processing often corresponds to a MF for the considered receive signals, the finite length of the transform leads to unwanted side lobes in radar imaging. Ideally, an (infinite-length) FT of a single frequency oscillation corresponds to a single sharp peak or Dirac in the spectrum and does not generate any entry at other frequency components. However, since radar signals are finite in time, frequency and space, the assumption of infinite length does not apply in reality. The received signal of a frequency ramp, symbol or the entire signal $y[k]$ can be interpreted as a snippet of a periodic signal $\bar{y}[k]$ of infinite length; what mathematically corresponds to a multiplication with a rectangular function. In discrete representation, this means

$$y[k] = \bar{y}[k] \cdot w[k] \tag{5.4}$$

for a window function $w[k]$, where $w[k]=0, \forall k \notin [k_0 - K/2, k_0 + K/2]$. The rectangular function

$$\text{rect}\left[\frac{k - k_0}{K}\right] = \begin{cases} 1, & \text{if } k \in [k_0 - \frac{K}{2}, k_0 + \frac{K}{2}] \\ 0, & \text{otherwise} \end{cases} \tag{5.5}$$

is the simplest window function $w[k]$. It is equal to 1 for an interval of K samples and is 0 outside this range. The FT of the signal (5.4) is the convolution

$$\mathcal{F}\{y[k]\} = \mathcal{F}\{\bar{y}[k] \cdot w[k]\} = \mathcal{F}\{\bar{y}[k]\} * \mathcal{F}\{w[k]\}. \tag{5.6}$$

If $w[k]$ equals the rectangular window (5.5) where $k_0 = 0$ and $\bar{y}[k] = \exp\{j2\pi \frac{v_0 k}{K}\}$ is an ideal oscillation whose normalized frequency v_0 is normalized to the resolution of the FT, (5.6) yields

$$\mathcal{F}\{y[k]\} = \delta[v - v_0] * \text{sinc}[v] = \text{sinc}[v - v_0]. \tag{5.7}$$

5.2 Windowing

This causes interferences in the spectrum at the coefficients $v \neq v_0$, $v \in \mathbb{Z}$, especially at the local extrema of the sinc function ($v \approx i + 0.5$, $i \in \mathbb{Z} \setminus \{0\}$), as can be observed in Fig. 5.4a. These interferences, which decrease periodically with distance to v_0, are referred to as side lobes. The closest and at the same time highest side lobe to the main maximum has a side lobe level of 13 dB. In the case of multiple targets, this level determines the maximum target dynamics of the radar, since in the case of targets of varying power in the signal, the weakest target can only be detected if it is higher than the highest side lobe of the strongest target. Conversely, this means that the weakest target must not be weaker than the strongest by more than 13 dB. If, for example, one considers the differences in RCS found in road traffic (see Table 1.2), this limitation leads to difficulties in the reliable detection of weakly reflecting road users such as pedestrians, especially in the vicinity of larger, strongly reflecting objects such as motor vehicles. To solve this target dynamic issue, special window functions are used to reduce the negative effects of finite FT and achieve a more favorable side lobe level. Two of the best known and most popular representatives in the radar field are the *Von Hann window* and the *Chebyshev window*.

The Von Hann window achieves a side lobe level suppression of 32 dB for the highest side lobe. The weight function

$$w_{\text{hann}}[k] = \frac{1}{2}\left[1 - \cos\left(\frac{2\pi k}{N-1}\right)\right] \tag{5.8}$$

and the associated spectrum $w_{\text{hann}}[v] = \mathcal{F}\{w_{\text{hann}}[k]\}$ of the Von Hann window are displayed in Fig. 5.4b. Compared to the rectangular window shown in Fig. 5.4a, the side lobe level is significantly reduced. However, the windowing causes a broadening of the main lobe. Compared to the rectangular window, which has a 3 dB lobe width of $0.89v$, the width of the main lobe of the Von Hann window is $1.44v$. A broadening and thus deterioration of the separation ability is a typical side effect of window functions, which must be taken into account when selecting the window function for the respective application; but which is usually accepted in favor of an improvement in the side lobe level.

The Chebyshev window (also called Dolph–Chebyshev window) provides a defined side lobe level, which is constant for all frequencies. The weights $w_{\text{cheb}}[k]$ in the time domain are determined by generating the desired amplitude distribution using Chebyshev polynomials in the frequency domain and subsequent transformation into the time domain [1]. The discrete amplitude distribution for a Chebyshev window of length $N+1$ in the frequency domain is defined as

$$w_0[v] = \frac{T_N\left(\beta \cos\left(\frac{\pi v}{N+1}\right)\right)}{T_N(\beta)}, \quad 0 \le v \le N. \tag{5.9}$$

$T_N(x)$ are Chebyshev polynomials of Nth order and first kind:

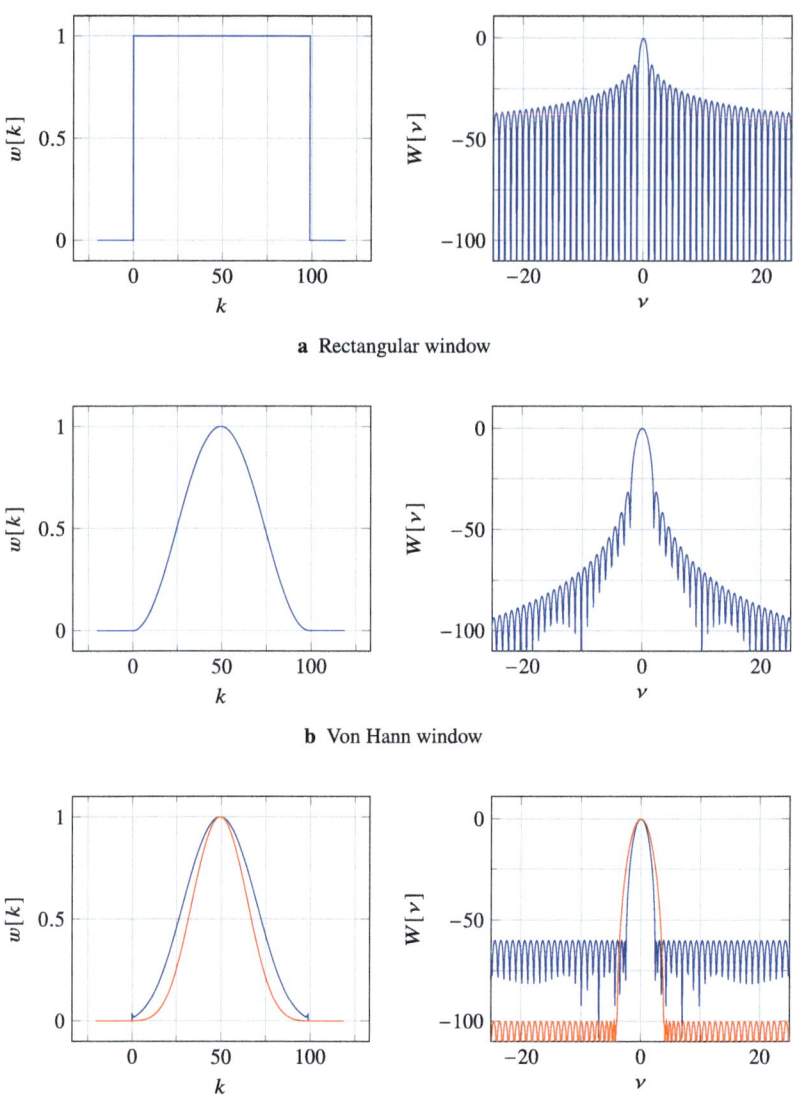

a Rectangular window

b Von Hann window

c Chebyshev window with 60 dB (*blue*) and 100 dB (*red*) side lobe level

Fig. 5.4 Common window functions in radar signal processing: **a** The rectangular window applies equal weighting to all samples in the time domain (*left*). In the spectrum (*right*), the maximum side lobe level is about 13 dB. **b** The Von Hann window is bell-shaped in the time domain (*left*), so that the samples are weighted less the closer they are to the beginning or end of the signal. In the spectrum (*right*) the maximum side lobe level is about 32 dB, however, the main beam is broadened compared to the rectangular window. **c** The Chebyshev window function is also bell-shaped in the time domain (*left*). Depending on the factor β in (5.9), a desired maximum side lobe level is obtained in the spectrum (*right*). Here, for example, the Chebyshev window functions with a level of 60 dB (*blue*) and 100 dB (*red*) are shown. Improving the side lobe level leads to a narrower bell-shaped time domain function and a broader main beam in the spectrum

Table 5.1 Side lobe level and main lobe width of different window functions

Window type	Side lobe level	3 dB main lobe width
Rectangular	13 dB	0.89ν
Von Hann	32 dB	1.44ν
Chebyshev ($\alpha = 60$ dB)	60 dB	1.44ν
Chebyshev ($\alpha = 100$ dB)	100 dB	1.84ν

$$T_N(x) = \begin{cases} \cos(N\cos^{-1}(x)), & \text{if } -1 \leq x \leq 1 \\ \cosh(N\cosh^{-1}(x)), & \text{if } x \geq 1 \\ (-1)^N \cosh(n\cosh^{-1}(-x)), & \text{if } x \leq -1 \end{cases} \quad (5.10)$$

The desired side lobe level can be set by adjusting the factor β in (5.9). For a desired side lobe level α_{dB} in decibels, the factor

$$\beta = \cosh\left(\frac{1}{N}\cosh^{-1}\left(10^{\alpha_{\text{dB}}/20}\right)\right) \quad (5.11)$$

must be used. The weights of the window function in the time domain are determined by means of a Fourier transform of the coefficients $w_0[\nu]$:

$$w_{\text{cheb}}[k] = \frac{1}{N+1} \sum_{\nu=0}^{N} W_0[\nu] e^{j2\pi \frac{k\nu}{N}}. \quad (5.12)$$

The resulting weights and the associated spectrum of the Chebyshev window are shown in Fig. 5.5c for the side lobe levels of 60 and 100 dB. Compared to the rectangular and Von Hann windows in Fig. 5.4a and b, respectively, the side lobes of the Chebyshev window are consistently flat and do not decrease with increasing distance from the main lobe. Furthermore, it can be observed that with a decreasing side lobe level α_{dB}, the main lobe broadens. A comparison of the side lobe levels and main lobe widths achieved by the three window types is summarized in Table 5.1.

5.3 Range and Velocity Estimation

The evaluation of the range and velocity information from the radar signals is discussed in this section. The modulation schemes introduced in Chap. 4 are revisited and the individual processing steps for each of them are discussed.

5.3.1 Evaluation of CW Radars

The receive signal of the CW radar from (4.6)

$$y_B(t) = A_{\text{rx}} \cos\left(2\pi\left(f_D t - \frac{2R f_c}{c}\right)\right)$$

is to be evaluated next with regard to the contained velocity information. As shown in Chap. 4, the range, velocity and angle characteristics of targets influence the phase of the received signal to be evaluated. The Doppler frequency f_D defines the frequency, or in the case of multiple targets, the frequency components, of the beat signal. The Doppler frequency f_D is therefore determined from the beat signal by a spectral analysis using a DFT, which corresponds to the MF for such a signal. First, the continuous-time signal is sampled at the sampling frequency f_s, which yields the discrete-time beat signal

$$y_B[k] = A_{\text{rx}} e^{j2\pi\left(\frac{f_D}{f_s}k - \frac{2R f_c}{c}\right)}. \tag{5.13}$$

The sampling frequency determines the maximum unambiguous Doppler frequency that can be evaluated. Therefore, the sampling frequency f_s should be chosen appropriately to avoid ambiguities. For a total duration of the received signal or observation time of the target of T_{obs}, a discrete-time signal with $K = f_s T_{\text{obs}}$ samples is obtained. The discrete-time signal (5.13) is multiplied by a window function $w[k]$ and then a DFT is performed along k. This yields the velocity spectrum

$$\mathcal{V}[\eta] = \sum_{k=0}^{K-1} w[k] y_B[k] e^{-j2\pi \frac{k\eta}{K}} \tag{5.14a}$$

$$= A_{\text{rx}} \sum_{k=0}^{K-1} w[k] e^{j2\pi\left(\frac{f_D}{f_s}k - \frac{2R f_c}{c}\right)} e^{-j2\pi \frac{k\eta}{K}} \tag{5.14b}$$

$$= \underbrace{A_{\text{rx}} e^{-j2\pi \frac{2R f_c}{c}}}_{A_v} \sum_{k=0}^{K-1} w[k] e^{-j2\pi \frac{k}{K}\left(\eta - f_D T_{\text{obs}}\right)} \tag{5.14c}$$

$$= A_v K \cdot W[\eta - f_D T_{\text{obs}}] \tag{5.14d}$$

of the beat signal (5.13). The velocity spectrum $\mathcal{V}[\eta]$ is also referred to as velocity profile. In contrast to the general discussion in Sect. 3.3.2, a monotone signal has not the shape of a sinc function in the spectrum, but the shape of the Fourier transform of the selected window function $w[\eta]$. The monotone signal is also shifted by $f_D T_{\text{obs}}$ in the spectrum. The spectrum of the window function in (5.14d) shows a maximum at the coefficient

$$\eta_0 = f_D T_{\text{obs}}. \tag{5.15}$$

5.3 Range and Velocity Estimation

From this, the corresponding Doppler frequency and consequently the velocity can be determined using the relationship from (1.23):

$$v = -\frac{c f_D}{2 f_c} = -\eta_0 \frac{c}{2 f_c T_{\text{obs}}}. \tag{5.16}$$

This also yields the Doppler or velocity resolution

$$\Delta v = \frac{c}{2 f_c T_{\text{obs}}} \tag{5.17}$$

in accordance to (3.17). The maximum unambiguous velocity is

$$v_{\text{ua}} = \left\{ -\frac{K}{2} \Delta v, \; \left(\frac{K}{2} - 1\right) \Delta v \right\} \stackrel{(3.36)}{\approx} \pm \frac{K}{2} \Delta v = \pm \frac{\lambda_c}{4} f_s \tag{5.18}$$

and is determined by the length of the DFT as introduced by (3.35). By applying the approximation (3.36) from Sect. 3.6, v_{ua} is defined only by the sampling frequency f_s and the wavelength of the carrier.

Theoretically, it is possible to evaluate the target range by evaluating the phase, as shown in Sect. 3.3.2. However, this is rarely used today due to the very low unambiguous range of $\lambda_c/2$. Also, the use of multiple frequency steps to improve the maximum unambiguous range is hardly used anymore. Instead, modulated signals such as FMCW are used.

5.3.2 Evaluation of FMCW Radars

The estimation of the velocities and ranges of targets by an FMCW radar is accomplished by extracting the frequencies contained in the beat signal

$$y_B(t) = A_{\text{rx}} \cos\left(\phi_B(t)\right) \tag{5.19a}$$

$$= A_{\text{rx}} \cos\left(2\pi \left(f_c \frac{2R}{c} + \frac{2v f_c}{c} t \pm \frac{B}{T} \frac{2R}{c} t\right)\right) \tag{5.19b}$$

of a single frequency ramp. In the first step, the signal is sampled at the sampling rate f_s, resulting in the discrete-time beat signal

$$y_B[k] = A_{\text{rx}} e^{j 2\pi \left(f_c \frac{2R}{c} + \left(f_D \pm \frac{B}{T} \frac{2R}{c}\right) \frac{k}{f_s}\right)}. \tag{5.20}$$

The signal is then multiplied by a window function and the range spectrum

$$\mathcal{R}[v] = \sum_{k=0}^{K-1} w[k] y_B[k] e^{-j2\pi \frac{kv}{K}} \tag{5.21a}$$

$$= A_{\text{rx}} \sum_{k=0}^{K-1} w[k] e^{j2\pi \left(f_c \frac{2R}{c} + \left(f_D \pm \frac{B}{T} \frac{2R}{c} \right) \frac{k}{f_s} \right)} e^{-j2\pi \frac{kv}{K}} \tag{5.21b}$$

is determined via a DFT. The number of samples $K = f_s T_{\text{obs}}$ is given by the sampling rate and observation time. If a single ramp is evaluated for FMCW, the observation time corresponds to the ramp duration $T_{\text{obs}} = T$. The spectrum thus gives

$$\mathcal{R}[v] = \underbrace{A_{\text{rx}} e^{j2\pi f_c \frac{2R}{c}}}_{A_R} \sum_{k=0}^{K-1} w[k] e^{-j2\pi \frac{k}{K} \left(v - \left(f_D T \pm \frac{2RB}{c} \right) \right)} \tag{5.21c}$$

$$= A_R K \cdot W \left[v - \left(f_D T \pm \frac{2RB}{c} \right) \right]. \tag{5.21d}$$

For a target, a maximum is detected at the index

$$v_0 = f_D T \pm \frac{2RB}{c} \tag{5.22}$$

in the spectrum with the corresponding beat frequency

$$\frac{v_0}{T} = f_D \pm \frac{2RB}{Tc}. \tag{5.23}$$

As mentioned in Sect. 4.1.2, the beat frequency contains both a velocity-dependent and a range-dependent component, which cannot be separated from each other when evaluating only a single frequency ramp. To distinguish the components from each other, a second frequency ramp with an inverted slope is evaluated, as described in Sect. 4.1.2. That is, if the first ramp is a rising frequency ramp, the second ramp should be a falling frequency ramp. Depending on whether the ramp is rising or falling, (5.23) results in the sum or the difference of the two frequency components. Subsequently, the system of equations (4.28) and (4.29) must be solved. This yields the velocity component

$$f_v = f_D \tag{5.24}$$

and the range component

$$f_R = \frac{2RB}{Tc} \tag{5.25}$$

5.3 Range and Velocity Estimation

of the beat frequency. These can now be used to determine the target velocity

$$v = -\frac{f_v c}{2 f_c} \tag{5.26}$$

and the target range

$$R = \frac{f_R T c}{2B}. \tag{5.27}$$

5.3.3 Evaluation of Chirp Sequence Radars

Unlike the previously considered radar types CW and FMCW, chirp sequence radars can determine range and velocity information unambiguously and separately using two Fourier transforms. First, the received signal or the beat signal of a single ramp l of the chirp sequence radar is considered, analogous to (4.33), (4.34) and (4.36):

$$y_B(l, t) = A_{rx} e^{j2\pi \left(f_c \frac{2R}{c} + f_D l T_{rep} \pm \frac{B}{T} \frac{2R}{c} t \right)}. \tag{5.28}$$

In the case of rising ramps, the last sign in (5.28) is a plus sign (+). The case of a falling ramp is analogous, only the sign changes to a minus sign (−). In the following, rising ramps (+) are assumed; the consideration for falling ramps (−) is analogous.

The individual ramps are sampled at a rate f_s. This results in the discrete-time beat signal

$$y_B[l, k] = A_{rx} e^{j2\pi \left(f_c \frac{2R}{c} + \frac{2 v f_c}{c} l T_{rep} + \frac{B}{T f_s} \frac{2R}{c} k \right)}. \tag{5.29}$$

The samples k of a single frequency ramp are referred to as the fast time and the intervals $l T_{rep}$ from ramp to ramp with $l = 0, \ldots, L-1$ are referred to as the slow time. If the sampled ramps of a chirp sequence frame are grouped row by row as a matrix, a matrix representation of the received signal analogous to Fig. 3.1a is obtained. A cut along a column corresponds to the slow time and a cut along a row to the fast time. The entry of the l-th row and k-th column of the frame matrix corresponds to $y_B[l,k]$ in (5.29). Since the velocity information depends only on the index l, a velocity-dependent phase change can only be detected along the columns of the frame matrix. The same applies to the range information: R only depends on the index k, which is why a range-dependent behavior can only be observed along the rows of the frame matrix. The two target parameters (range and velocity) are therefore independent of each other with respect to the indices k and l. For this reason, velocity and range can be determined independently of each other by a DFT along the columns, i.e. the fast time k, and by a DFT along the rows, i.e. the slow time l.

5.3.3.1 Range Estimation

To evaluate the range information, i.e. to obtain the range profile, a DFT is performed along the fast time k of the beat signal $y_B[l,k]$ (5.29) of a single ramp l:

$$\mathcal{R}[l,\nu] = \sum_{k=0}^{K-1} w[k] y_B[l,k] e^{-j2\pi \frac{k\nu}{K}} \tag{5.30a}$$

$$= A_{rx} \sum_{k=0}^{K-1} w[k] e^{j2\pi \left(f_c \frac{2R}{c} + f_D l T_{rep} + \frac{B}{T f_s} \frac{2R}{c} k \right)} e^{-j2\pi \frac{k\nu}{K}}. \tag{5.30b}$$

Only the components that depend on index k are relevant for the evaluation of the DFT. Rearranging the equation yields

$$\mathcal{R}[l,\nu] = \underbrace{A_{rx} e^{j2\pi f_c \frac{2R}{c}}}_{A_0} e^{j2\pi f_D l T_{rep}} \sum_{k=0}^{K-1} w[k] e^{-j2\pi \frac{k}{K} \left(\nu - \frac{2BR}{c} \right)} \tag{5.30c}$$

$$= A_0 e^{j2\pi f_D l T_{rep}} \cdot K \cdot W\left[\nu - \frac{2BR}{c} \right]. \tag{5.30d}$$

The window function $w[k]$ has a width of K, that is $w[k] \neq 0, \forall k \in [0, K)$ and 0 otherwise. Except for a difference in phase due to $\exp\{j2\pi f_D l T_{rep}\}$, the range profiles for the individual ramps are identical. For a target, the spectrum (5.30d) shows a peak at the coefficient

$$\nu_0 = \frac{2BR}{c}. \tag{5.31}$$

Solving this equation for the range and taking into account the minimal change of ν by one provides the range resolution

$$R = \nu_0 \underbrace{\frac{c}{2B}}_{\Delta R} \quad \Rightarrow \quad \Delta R = \frac{c}{2B}. \tag{5.32}$$

The maximum unambiguously detectable range, limited by the DFT, is given by

$$R_{ua} = (K-1)\Delta R \approx f_s T \frac{c}{2B}. \tag{5.33}$$

5.3.3.2 Velocity Estimation

To evaluate the velocity information, a DFT is performed along the slow time of any sample k along all ramps of the frame:

5.3 Range and Velocity Estimation

$$\mathcal{V}[\eta, k] = \sum_{l=0}^{L-1} w[l] y_B[l, k] e^{-j2\pi \frac{l\eta}{L}} \tag{5.34a}$$

$$= A_{rx} \sum_{l=0}^{L-1} w[l] e^{j2\pi \left(f_c \frac{2R}{c} + f_D l T_{rep} + \frac{B}{Tf_s} \frac{2R}{c} k \right)} e^{-j2\pi \frac{l\eta}{L}}. \tag{5.34b}$$

Only the components that depend on index l are relevant for the evaluation of the DFT. Rearranging the equation results in

$$\mathcal{V}[\eta, k] = \underbrace{A_{rx} e^{j2\pi f_c \frac{2R}{c}}}_{A_0} e^{j2\pi \frac{B}{Tf_s} \frac{2R}{c} k} \sum_{l=0}^{L-1} w[l] e^{-j2\pi \frac{l}{L} \left(\eta - f_D L T_{rep} \right)} \tag{5.34c}$$

$$= A_0 e^{j2\pi \frac{B}{Tf_s} \frac{2R}{c} k} \cdot L \cdot W\left[\eta - f_D T_{obs} \right]. \tag{5.34d}$$

In order to detect and distinguish both positive and negative Doppler shifts, the DFT coefficients are evaluated in the range of $\eta \in [-L/2, L/2-1]$. Here, negative indices correspond to negative Doppler shifts and thus positive relative velocities v; the opposite applies to positive indices. The window function $w[l]$ applied has the width L. Analogous to the evaluation of the range, the velocity profiles (5.34d) for the evaluation along other indices k are almost identical. Only the phase differs due to the term $\exp\{j2\pi \frac{B}{Tf_s} \frac{2R}{c} k\}$.

For a target, the velocity spectrum (5.34d) shows a peak at index

$$\eta_0 = f_D T_{obs} = -\frac{2 f_c v T_{obs}}{c}. \tag{5.35}$$

Solving the equation for the Doppler frequency or velocity yields the achievable Doppler or velocity resolution with a minimum increment of η of one:

$$\Delta f_D = \frac{1}{T_{obs}} = \frac{1}{LT} \quad \text{and} \quad \Delta v = \frac{c}{2 f_c T_{obs}} = \frac{c}{2 f_c LT}. \tag{5.36}$$

The maximum unambiguously detectable Doppler frequency and velocity are determined by the resolution and length of the DFT by

$$f_{D,ua} = \left\{ -\Delta \frac{L}{2} f_D, \left(\frac{L}{2} - 1 \right) \Delta f_D \right\} \stackrel{(3.36)}{\approx} \pm \frac{L}{2} \Delta f_D = \pm \frac{1}{2T} \tag{5.37}$$

and

$$v_{ua} = \left\{ -\frac{L}{2} \Delta v, \left(\frac{L}{2} - 1 \right) \Delta v \right\} \stackrel{(3.36)}{\approx} \pm \frac{L}{2} \Delta v = \pm \frac{\lambda_c}{4T}. \tag{5.38}$$

The approximation (3.36) from Sect. 3.6 shows that v_{ua} is characterized by the ramp duration T and the wavelength of the carrier.

5.3.3.3 Joint Estimation of Range and Velocity

An isolated evaluation of range and velocity, as described in the previous Sects. 5.3.3.1 and 5.3.3.2, does not allow the different detected velocities and ranges to be combined in the case of multiple targets. This is only possible if the parameters are not evaluated separately, but jointly by means of a two-dimensional DFT. The result of this evaluation is referred to as a range-Doppler matrix or R-v matrix. If the received signal is considered to be a matrix of size $L \times K$ with entries $y_B[l,k]$ (5.29), the columns contain the fast-time samples of the ramps and the rows contain the individual ramps. Thus, velocity and range evaluation corresponds to a column-wise and row-wise DFT of this matrix. The result of this 2D DFT is again a matrix of size $L \times K$, where the columns correspond to the range indices v and the rows to the velocity indices η, as illustrated in Fig. 3.1. Since the two DFT operations are linear, the order of evaluation is interchangeable. However, often first the DFT along the fast time, i.e., the range DFT of each individual ramp, is determined. The reason for this is that only the samples of the l-th ramp are needed for the l-th range profile. This means that the evaluation of the first ramps can begin even before the last ramp of the frame has been sent. For the velocity DFT, on the other hand, it is necessary to first record all the measured samples in a frame. The following discussion is therefore limited to the former order.

First, a range evaluation of the beat signal as in (5.30) is performed for each ramp. The result can be represented as a matrix whose rows contain the range profiles of the ramps. Subsequently, another DFT is performed along the columns of this matrix. The evaluation is similar to (5.34), but instead of the beat signal, the columns of the matrix of the range evaluation are used. Combining result in the R-v matrix of the chirp sequence measurement:

$$\mathcal{I}[\eta, v] = \sum_{l=0}^{L-1} \sum_{k=0}^{K-1} \underbrace{w[l]w[k]}_{w[l,k]} y_B[l,k] e^{-j2\pi \frac{kv}{K}} e^{-j2\pi \frac{l\eta}{L}} \qquad (5.39a)$$

$$= \underbrace{\sum_{l=0}^{L-1} A_0 e^{j2\pi f_D l T_{\text{rep}}} \cdot K \cdot W\left[l, v - \frac{2BR}{c}\right] e^{-j2\pi \frac{l\eta}{L}}}_{\text{range DFT (5.30d)}} \qquad (5.39b)$$

$$= A_0 \cdot K \cdot \underbrace{\sum_{l=0}^{L-1} W\left[l, v - \frac{2BR}{c}\right] e^{-j2\pi \frac{l}{L}(\eta - f_D L T_{\text{rep}})}}_{\text{Doppler DFT (5.30d)}} \qquad (5.39c)$$

$$= A_0 \cdot K \cdot L \cdot W\left[\eta - f_D T_{\text{obs}}, v - \frac{2BR}{c}\right]. \qquad (5.39d)$$

The product $w[l] \cdot w[k]$ of the window functions can be summarized to a composite two-dimensional window function $w[l,k]$. This two-dimensional discrete window can be represented in matrix form, just like the discrete-time beat signal. The window matrix has the size $L \times K$ and entries $w[l,k]$. The product $K \cdot L$ in (5.39d) corresponds

5.3 Range and Velocity Estimation

to the integration gain I achieved by the two DFTs. For each measured target, the R-v matrix has a peak at the corresponding entry $[\eta_0, \nu_0]$ analogous to (5.31) and (5.35). This means that the ranges and velocities of all targets can be unambiguously associated even in the case of multiple targets.

5.3.3.4 Example of a Range Measurement with a Chirp Sequence Radar

Figure 5.5 shows a measurement of a single target at 5.3 m in an anechoic chamber. The length of the anechoic chamber is about 6.5 m whereby the signal level in the range profile for ranges of $R > 6.5$ m is only defined by the intrinsic noise of the system. Signal responses in the range of $R < 6.5$ m contain various reflections from within the anechoic chamber. The responses at very short ranges are caused by direct crosstalk and reflections between the transmit and receive channels. The used radar parameters for the chirp sequence signal are summarized in Table 5.2.

5.3.4 Evaluation of OFDM Radars

Like the chirp sequence radar, the OFDM radar allows the range and velocity to be estimated from different measurement dimensions using two one-dimensional or one two-dimensional Fourier transform. Unlike with a chirp sequence radar, however, the

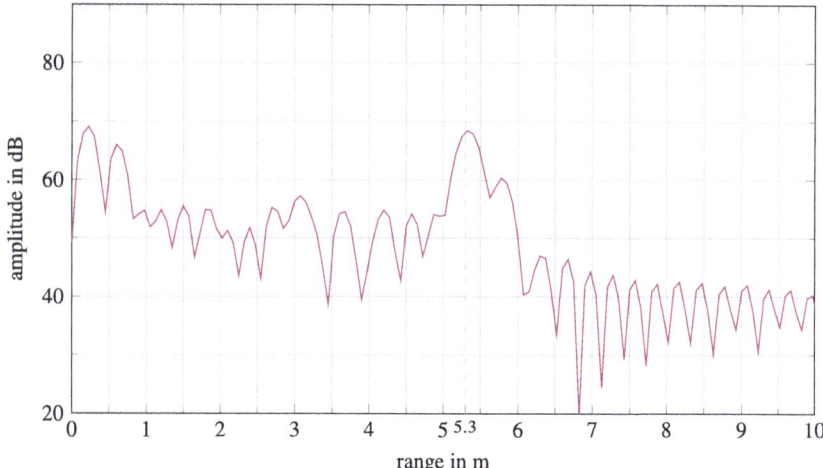

Fig. 5.5 Range measurement with a chirp sequence signal for a target at 5.3 m. The used radar parameters for the chirp sequence signal are summarized in Table 5.2. The spikes at very short ranges are caused by crosstalk between the transmit and receive channels

Table 5.2 Radar parameters and corresponding range resolution of the chirp sequence radar system used for the measurement in Fig. 5.5

Modulation parameter		
Carrier frequency	f_c	77 GHz
Bandwidth	B	512 MHz
Ramp duration	T	30 μs
Number of chirps	L	256
Performance parameter		
Range resolution	ΔR	0.3 m

received signal is not evaluated directly in slow time and fast time, but in the so-called time-frequency domain, also known as a *spectrogram*, which is achieved by a DFT along the fast time, see (4.53) in Sect. 4.2.1. Subsequently, the remaining modulation is removed (see (4.54)), resulting in the channel spectrogram **D** (4.55) with $L \times N$ elements

$$D[l,n] = A_{\mathrm{rx}} \underbrace{e^{-j2\pi f_c \frac{2R}{c}}}_{\text{const.}} \underbrace{e^{-j2\pi n \Delta f \frac{2R}{c}}}_{\propto n} \underbrace{e^{j2\pi f_D l T_{\mathrm{OFDM}}}}_{\propto l} . \tag{5.40}$$

The rows of the matrix **D** thus correspond to the slow time, i.e. the OFDM symbols, and the columns to the subcarriers of the frame. A range-proportional behavior can only be observed along the subcarriers, and a velocity-proportional behavior can only be observed along the slow time, i.e. OFDM symbols. There is also a constant phase component in (5.40), which only depends on the target range but is identical for all subcarriers and symbols.

5.3.4.1 Range Estimation

Since the range information is evaluated symbol by symbol across the subcarriers, the processing can already begin while the measurement is still continuing. To determine the range profile and thus the range R of the target, an IDFT is performed for each symbol l across its subcarriers n:

$$\mathcal{R}[l, v] = \sum_{n=0}^{N-1} w[n] D[l,n] e^{j2\pi \frac{nv}{N}} \tag{5.41a}$$

$$= \underbrace{A_{\mathrm{rx}} e^{-j2\pi \frac{2R}{c}}}_{A_0} \sum_{n=0}^{N-1} w[n] e^{-j2\pi n \Delta f \frac{2R}{c}} e^{j2\pi f_D l T_{\mathrm{OFDM}}} e^{j2\pi \frac{nv}{N}} . \tag{5.41b}$$

Only the components that depend on index n are relevant for the evaluation of the IDFT. Rearranging the equation yields

5.3 Range and Velocity Estimation

$$\mathcal{R}[l, v] = A_0 e^{j2\pi f_D/T_{\text{OFDM}}} \sum_{n=0}^{N-1} w[n] e^{j2\pi \frac{n}{N}\left(v - \frac{2\Delta f N R}{c}\right)} \quad (5.41c)$$

$$= A_0 e^{j2\pi f_D/T_{\text{OFDM}}} \cdot N \cdot W\left[v - \frac{2BR}{c}\right]. \quad (5.41d)$$

Comparing (5.41d) with the range profile of the chirp sequence evaluation (5.30d) shows a very similar result despite the significantly different modulation and the different approach in the evaluation. The window function $w[n]$ in (5.41a) has the width N. The range spectrum has a peak for each target at the corresponding range R for the respective index

$$v_0 = \frac{2BR}{c}. \quad (5.42)$$

Solving this equation for the range R and taking into account the minimal change of v by one gives the range resolution

$$\Delta R = \frac{c}{2B}. \quad (5.43)$$

The length N of the IDFT determines the maximum unambiguous range

$$R_{\text{ua}} = (N-1)\Delta R = \frac{c(N-1)}{2B} \approx \frac{c}{2\Delta f} = \frac{cT}{2}. \quad (5.44)$$

The maximum unambiguous range for OFDM is therefore defined by the duration of the OFDM symbol. Unlike other radar modulations that use repetitive fundamental waveforms within a measurement, such as chirp sequence radar, conventional OFDM modulation experiences no ambiguities for longer ranges. The reason for this is that different modulation phases are usually selected for the L OFDM symbols of the frame. If long ranges $R > R_{\text{ua}}$ result in propagation times $\tau > T$, the respective l-th received symbol is superimposed on the $(l+1)$-th transmitted symbol at the receiver, as shown in Fig. 4.14 in Sect. 4.2.1. In the demodulation step (4.54), symbol d_{ln} is consequently divided by symbol $d_{(l+1)n}$, which does not resolve the modulation but instead creates a random phase that is different for each $D[l,n]$. As a result, the phases in the IDFT do not correlate; no peak arises in the spectrum $\mathcal{R}[l,v]$ and thus no ambiguities.

In addition to the maximum unambiguous range, OFDM features a maximum range $R_{\text{max}} < R_{\text{ua}}$, which is defined by the length of the cyclic prefix:

$$R_{\text{max}} = N_{\text{cp}}\Delta R = \frac{cT_{\text{cp}}}{2}. \quad (5.45)$$

This defines the maximum propagation delay up to which the n-th receive symbol completely overlaps with the n-th transmit symbol in the range from T_{cp} to T_{OFDM}. As a result, the entire symbol duration T contributes to the evaluation and thus to the integration gain in the IDFT. If the delay τ of the receive signal is greater than T_{cp} but less than the symbol duration, i.e. $T_{cp} < \tau < T_{OFDM}$, the integration gain is reduced in proportion to $(\tau - T_{cp})/T$.

5.3.4.2 Velocity Estimation

With OFDM, the velocity is estimated by a DFT along the OFDM symbols l (slow time) of the time-frequency matrix **D**:

$$\mathcal{V}[\eta, n] = \sum_{l=0}^{L-1} w[l] D[l, n] e^{-j2\pi \frac{l\eta}{L}} \tag{5.46a}$$

$$= \underbrace{A_{rx} e^{-j2\pi \frac{2R}{c}}}_{A_0} \sum_{l=0}^{L-1} w[l] e^{-j2\pi n \Delta f \frac{2R}{c}} e^{j2\pi f_D l T_{OFDM}} e^{-j2\pi \frac{l\eta}{L}}. \tag{5.46b}$$

Only the components that depend on index l are relevant for the evaluation of the DFT. Rearranging the equation results in

$$\mathcal{V}[\eta, n] = A_0 e^{-j2\pi n \Delta f \frac{2R}{c}} \sum_{l=0}^{L-1} w[l] e^{-j2\pi \frac{l}{L}(\eta - f_D L T_{OFDM})} \tag{5.46c}$$

$$= A_0 e^{-j2\pi n \Delta f \frac{2R}{c}} \cdot L \cdot W[\eta - f_D L T_{OFDM}]. \tag{5.46d}$$

In order to allow unambiguous acquisition of both negative and positive velocities, the DFT coefficients are evaluated in the range $\eta \in [-L/2, L/2-1]$—as in the chirp sequence evaluation. The velocity spectrum $\mathcal{V}[\eta,l]$ shows a peak at the index

$$\eta_0 = f_D L T_{OFDM} = -\frac{2 f_c v L T_{OFDM}}{c} \tag{5.47}$$

corresponding to the target velocity. If, analogous to the range evaluation, the minimal increment of η by one is considered, the resolutions for the Doppler frequency and the velocity are given by

$$\Delta f_D = \frac{1}{L T_{OFDM}} \quad \text{and} \quad \Delta v \frac{c}{2 f_c L T_{OFDM}}. \tag{5.48}$$

The length of the DFT provides the maximum unambiguous Doppler frequency and velocity of

$$f_{D,ua} = \left\{ -\frac{L}{2} \Delta f_D, \left(\frac{L}{2}-1\right) \Delta f_D \right\} \stackrel{(3.36)}{\approx} \pm \frac{L}{2} \Delta f_D = \pm \frac{1}{2 T_{OFDM}} \tag{5.49}$$

5.3 Range and Velocity Estimation

and

$$v_{ua} = \left\{ -\frac{L}{2}\Delta v, \left(\frac{L}{2}-1\right)\Delta v \right\} \stackrel{(3.36)}{\approx} \pm\frac{L}{2}\Delta v = \pm\frac{c}{4f_c T_{OFDM}}. \quad (5.50)$$

The approximation (3.36) from Sect. 3.6 shows that v_{ua} is characterized by the symbol duration T_{OFDM} and the wavelength of the carrier.

Similar to the range estimation, the velocity estimation for OFDM distinguishes between maximum unambiguous and maximum velocity. The reason for this is that the Doppler shift experienced by the received signal causes a shift of the OFDM spectrum (see Fig. 4.12). However, the signal is still evaluated at the original subcarrier frequencies $n\Delta f$. At these frequencies, the subcarriers are no longer orthogonal due to the shift of the spectrum. Due to this, there is a small amount of crosstalk of all subcarriers into the received signal of the n-th subcarrier; This is referred to as intercarrier interference (ICI). Additionally, there are losses in the signal amplitude because the evaluation is not carried out at the maximum of the spectrum (the sinc function), but rather slightly offset from it. The further the spectrum is shifted from the original frequency, the greater both losses in the velocity evaluation. For this reason, a maximum acceptable Doppler shift is defined, up to which the losses during the velocity evaluation are below a user-defined value, e.g. 1 dB. In most cases, the maximum ratio of subcarrier spacing to Doppler shift is derived from the maximum expected Doppler shift $f_{D,max}$ in the channel:

$$\beta = \left|\frac{\Delta f}{f_{D,max}}\right| \quad \text{where} \quad \beta \gg 1. \quad (5.51)$$

This defines the maximum velocity accepted by the system of

$$|v_{max}| = \frac{\Delta f \lambda}{2\beta}. \quad (5.52)$$

For a value of $\beta \approx 4$, this results in a loss of approximately 1 dB.

5.3.4.3 Joint Estimation of Range and Velocity

Analogous to chirp sequence radar, OFDM radar usually first determines the range evaluation of all OFDM symbols and then performs the velocity evaluation. This corresponds to a row-by-row IDFT of the time-frequency matrix **D**, followed by a column-by-column DFT of the thus resulting matrix. However, the order can also be reversed. Eventually, this results in the R-v matrix:

$$I[\eta, \nu] = \sum_{l=0}^{L-1}\sum_{n=0}^{N-1} \underbrace{w[l]w[n]}_{w[l,k]} D[l,n] e^{j2\pi \frac{n\nu}{K}} e^{-j2\pi \frac{l\eta}{L}} \quad (5.53a)$$

$$\underbrace{\phantom{\sum_{l=0}^{L-1}\sum_{n=0}^{N-1} w[l]w[n] D[l,n] e^{j2\pi \frac{n\nu}{K}} e^{-j2\pi \frac{l\eta}{L}}}}_{\text{range IDFT (5.41d)}}$$

$$= \sum_{l=0}^{L-1} A_0 e^{j2\pi f_D l T_{\text{OFDM}}} \cdot N \cdot W\left[\nu - \frac{2BR}{c}\right] e^{-j2\pi \frac{l\eta}{L}} \quad (5.53b)$$

$$= A_0 \cdot N \cdot \underbrace{\sum_{l=0}^{L-1} W\left[l, \nu - \frac{2BR}{c}\right] e^{j2\pi f_D l T_{\text{OFDM}}} e^{-j2\pi \frac{l\eta}{L}}}_{\text{Doppler DFT (5.46d)}} \quad (5.53c)$$

$$= A_0 \cdot N \cdot L \cdot W\left[\eta - f_D T_{\text{OFDM}}, \nu - \frac{2BR}{c}\right]. \quad (5.53d)$$

The total integration gain achieved by the OFDM range and velocity estimation is thus $I = N \cdot L$.

5.3.4.4 Example of a Range Measurement with an OFDM Radar

Figure 5.6 presents a measurement of a single target located at 5.3 m within an anechoic chamber. The chamber's length is approximately 6.5 m, meaning that for ranges greater than 6.5 m, the signal level in the range profile is influenced solely by the system's intrinsic noise. For ranges less than 6.5m, the signal responses exhibit

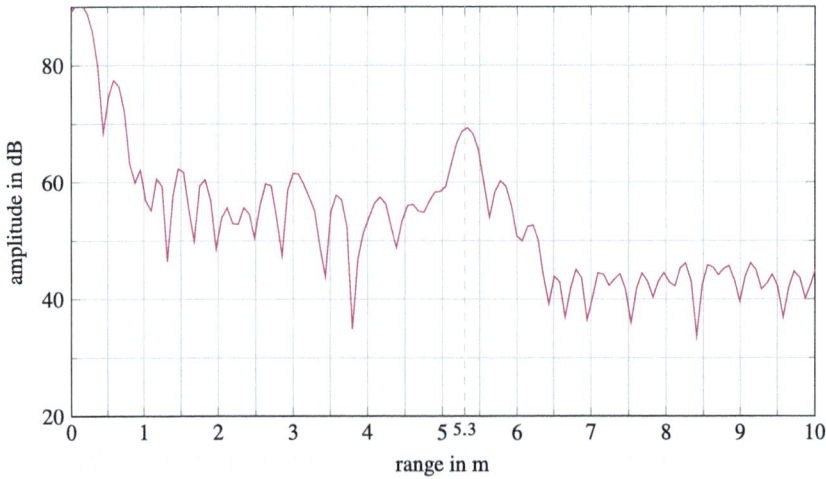

Fig. 5.6 Range measurement with an OFMD signal for a target at 5.3 m. The used radar parameters for the OFMD signal are summarized in Table 5.3. The spikes at very short ranges are caused by crosstalk between the transmit and receive channels

5.3 Range and Velocity Estimation

Table 5.3 Radar parameters and corresponding range performance of the OFMD radar system used for the measurement in Fig. 5.6

Modulation parameters		
Carrier frequency	f_c	77 GHz
Bandwidth	B	512 MHz
Subcarrier spacing	Δf	500 kHz
Number of subcarriers	N	1024
Symbol duration	T	2 μs
Number of symbols	L	2048
Cyclic prefix duration	T_{cp}	0.4 μs
Performance parameters		
Range resolution	ΔR	0.3 m
Maximum range	R_{ua}	60 m

multiple reflections within the chamber. The responses observed at very short ranges are primarily due to direct crosstalk and reflections between the transmitting and receiving channels. A summary of the radar parameters used for the OFDM signal can be found in Table 5.3.

5.3.5 Evaluation of PMCW Radars

A distinctive characteristic of the PMCW radar compared to the radar types considered so far is that the range evaluation is not achieved by evaluating the receive phase using a spectral analysis or a Fourier-based approach, but by correlation in the time domain. The velocity evaluation, on the other hand, is still achieved by a Fourier transform along L code sequences of the frame. Considering the discrete-time receive signal of the l-th code sequence

$$y[l, k] = A_{rx} x^{PN}\left[kT_s - \frac{2R}{c}\right] \cdot e^{j2\pi(f_c \frac{2R}{c} - f_D lT)}, \tag{5.54}$$

it can be represented as an $L \times K$ matrix with elements $y[l, k]$ by stacking the L receive signals row-wise. Here, too, the samples k (of a single code sequence) correspond to the fast time of the signal and a consideration of the columns to the slow time of the signal. Likewise, a range-dependent influence along the fast time and a velocity-dependent influence along the slow time can be observed.

5.3.5.1 Range Estimation via Correlation

With the PMCW radar, the target range is determined by comparing the received code sequence with the transmitted reference sequence. For a simpler representation, it is assumed from here on that the sampling interval T_s corresponds to the chip rate T_c of the sequence. In this case, $x^{\text{PN}}[kT_s]$ corresponds to the k-th symbol of the sequence, which is hereinafter referred to as $x^{\text{PN}}[kT]$ for the sake of simplicity. Thus, the correlation of the received and reference signals yields the range profile

$$\mathcal{R}[l, \nu] = \sum_{k=1}^{K} y_l^*[k] x^{\text{PN}}[k+\nu]. \tag{5.55}$$

This assumes a cyclic code sequence, i.e.

$$x^{\text{PN}}[k] = x^{\text{PN}}[k+K] \quad \text{and} \quad x^{\text{PN}}[-k] = x^{\text{PN}}[K-k].$$

Substituting (5.54) into the correlation (5.55) gives

$$\mathcal{R}[l, \nu] = A_{\text{rx}}^* \cdot e^{-j2\pi(f_c \frac{2R}{c} - f_D l T)} \underbrace{\sum_{k=1}^{K} \left(x^{\text{PN}}\left[k - \frac{2R}{cT_c} \right] \right)^* x^{\text{PN}}[k+\nu]}_{K \cdot \rho_{x^{\text{PN}}}\left[\nu - \frac{2R}{cT_c} \right]}, \tag{5.56}$$

where the normalized autocorrelation function

$$\rho_{x^{\text{PN}}}\left[\nu - \frac{2R}{cT_c} \right] = \frac{1}{K} \sum_{k=1}^{K} \left(x^{\text{PN}}\left[k - \frac{2R}{cT_c} \right] \right)^* x^{\text{PN}}[k+\nu] \tag{5.57}$$

is shifted by $2R/(cT_c)$. By choosing a suitable sequence with an optimal autocorrelation property, (5.57) shows a maximum at index

$$\nu_0 = \frac{2R}{cT_c}. \tag{5.58}$$

Figure 4.17 in Sect. 4.2.2 shows the autocorrelation functions of maximum length, Gold and Kasami sequences, which are all suitable for PMCW radars. In particular, maximum length sequences show a very high PSLR of $20 \log_{10}(K)$. Solving (5.58) for the range results in the resolution

$$\Delta R = \frac{cT_c}{2} = \frac{c}{2B}. \tag{5.59}$$

5.3 Range and Velocity Estimation

The resolution, as presented in the last expression, is identical to the range resolution for FMCW, chirp sequence, and OFDM. In the case of PMCW, the bandwidth is defined by the symbol rate, see (4.59). This in turn means that the correlation of the time signals has a maximum temporal resolution of T_c, which in turn determines the limit of the range resolution.

Since the measurement frame of a PMCW radar consists of a single code sequence that is repeated L times, a quasiperiodic signal with a period duration T is generated. For this reason, time delays at the receiver can only be unambiguously evaluated in this interval. Targets that produce a delay of $2R/c > T$ give rise to ambiguities in the evaluation, which should be taken into account when interpreting the signal. The maximum unambiguous range therefore is

$$R_{ua} = \frac{cT}{2}. \tag{5.60}$$

5.3.5.2 Velocity Estimation via DFT

Subsequent to the range evaluation, the velocity is estimated by a DFT along the slow time of each range index v of the range spectrum (5.56). This finally yields the R-v matrix of the PMCW measurement:

$$I[\eta, v] = \sum_{l=0}^{L-1} w[l] \mathcal{R}[l, v] e^{-j2\pi \frac{l\eta}{L}} \tag{5.61a}$$

$$= \underbrace{A_{rx}^* e^{-j2\pi f_c \frac{2R}{c}}}_{A_0^*} \cdot K \cdot \rho_{x^{PN}}\left[v - \frac{2R}{cT_c}\right] \cdot \sum_{l=0}^{L-1} w[l] e^{j2\pi f_D lT} e^{-j2\pi \frac{l\eta}{L}} \tag{5.61b}$$

$$= A_0^* \cdot K \cdot \rho_{x^{PN}}\left[v - \frac{2R}{cT_c}\right] \cdot \sum_{l=0}^{L-1} w[l] e^{-j2\pi \frac{l}{L}(\eta - f_D LT)} \tag{5.61c}$$

$$= A_0^* \cdot K \cdot \rho_{x^{PN}}\left[v - \frac{2R}{cT_c}\right] \cdot L \cdot W[\eta - f_D T_{obs}]. \tag{5.61d}$$

For PMCW, the observation time is given by $T_{obs} = LT$. In order to be able to capture both positive and negative velocities, the unambiguous range of values $\eta \in [-L/2, L/2 - 1]$ is evaluated. Negative indices correspond to negative Doppler shifts or positive velocities and vice versa. The velocity spectrum or R-v matrix $I[v,\eta]$ shows a maximum at the index

$$\eta_0 = f_D T_{obs} = -\frac{2 f_c v L T}{c}. \tag{5.62}$$

Solving this equation for the Doppler yields the Doppler and velocity resolutions

$$\Delta f_D = \frac{1}{LT} \quad \text{and} \quad \Delta v = \frac{c}{2 f_c LT}, \tag{5.63}$$

respectively. The maximum unambiguously evaluable Doppler frequency and velocity is defined by the respective resolution and the length of the DFT:

$$f_{D,ua} = \left\{ -\frac{L}{2} \Delta f_D, \left(\frac{L}{2} - 1 \right) \Delta f_D \right\} \stackrel{(3.36)}{\approx} \pm \frac{L}{2} \Delta f_D = \pm \frac{1}{2T} \tag{5.64}$$

and

$$v_{ua} = \left\{ -\frac{L}{2} \Delta v, \left(\frac{L}{2} - 1 \right) \Delta v \right\} \stackrel{(3.36)}{\approx} \pm \frac{L}{2} \Delta v = \pm \frac{c}{4 f_c T}. \tag{5.65}$$

The approximation (3.36) from Sect. 3.6 shows that v_{ua} is characterized by the code duration T and the wavelength of the carrier.

5.3.5.3 Example of a Range Measurement with a PMCW Radar

In Fig. 5.7, the measurement of a single target at a distance of 5.3 m inside an anechoic chamber is shown. The chamber itself is about 6.5 m long, which means that for ranges beyond 6.5 m, the signal level in the range profile is determined only by the system's inherent noise. For shorter ranges, reflections within the chamber contribute to the signal responses. At very short ranges, the observed responses are primarily caused

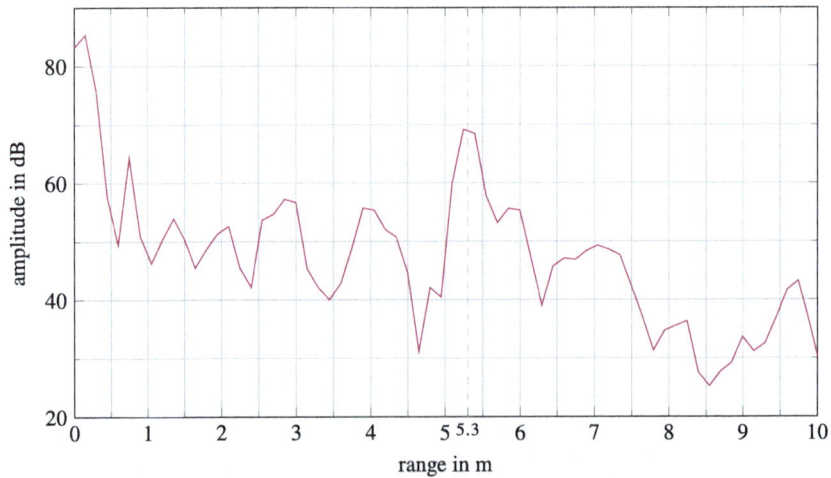

Fig. 5.7 Range measurement with a chirp sequence signal (maximum length sequence) for a target at 5.3 m. The used radar parameters for the PMCW signal are summarized in Table 5.4. The spikes at very short ranges are caused by crosstalk between the transmit and receive channels

Table 5.4 Radar parameters and corresponding range performance of the PMCW radar system used for the measurement in Fig. 5.7

Modulation parameters		
Carrier frequency	f_c	77 GHz
Bandwidth	B	500 MHz
Code duration	T	2 µs
Code length	K	1024
Number of repetitions	L	2048
Performance parameters		
Range resolution	ΔR	0.3 m
Maximum range	R_{ua}	300 m

by direct crosstalk and reflections between the transmitter and receiver channels. The radar parameters for the PMCW signal are summarized in Table 5.4.

5.4 Target Detection using CFAR Algorithms

After the DFT evaluation of the signal in one or more measurement dimensions (slow time, fast time, channels), the measured signal is represented in the image domain. Depending on the signal processing chain, this image corresponds to a parameter space with up to three dimensions: range, velocity and angle (see Fig. 5.1). As a result of the DFT or angle estimation, the target power is now concentrated in individual image cells, which correspond to the measured target ranges, velocities and angles. In contrast, noise and interference power are distributed rather evenly in the image, whereby targets stand out against the noise level and can usually be clearly recognized as peaks.

This is illustrated in Fig. 5.8 using the example of a range evaluation of a measurement with a chirp sequence radar. Several signal peaks at various ranges can be easily recognized, for example strong targets at about 9 and 11.3 m, as well as several other targets between 12 and 16 m. In addition, several weaker responses at some other ranges can be observed. The task of the CFAR algorithm is to decide which peaks are classified as targets and what is interpreted as noise. This is done by comparing the amplitudes with a threshold. To do so, first a threshold is established; then it is examined for each cell whether its amplitude exceeds this threshold or not. If it does, this cell is considered to contain a potential target. The simplest approach for this is to choose a fixed threshold for all cells, such as the mean value of all cells, as shown in Fig. 5.8 by the *red dashed* curve. However, the noise level of the measured data curve in Fig. 5.8 (*blue*) changes over range. One reason for this is the filter characteristics of the radar. Using a constant threshold therefore leads to a high error rate due to false detections and undetected, i.e. missed, targets.

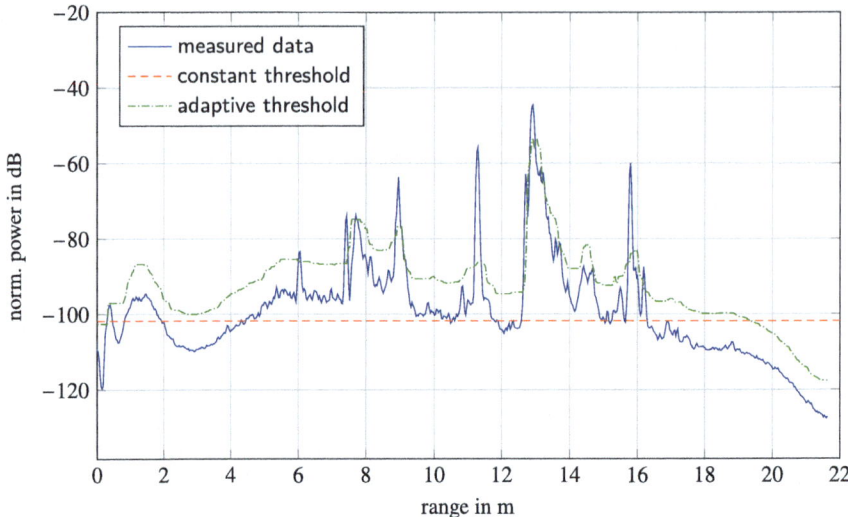

Fig. 5.8 Comparison of a constant (*red dashed*) and adaptive CFAR (*green dash-dotted*) threshold approach using the example of a range evaluation of a measurement with a chirp sequence radar (*blue*). The constant threshold yields a too low threshold where too much of the signal would pass the detection. The adaptive threshold follows the difference in the signal level caused by, e.g. non-constant noise or frequency-dependent filter characteristics as observable for the ranges from 0 to 4 m and from 18 to 20 m ; This leads to less erroneous detections

False detections or false alarms are detections by the target detection algorithm that do not correspond to actual targets. The reason for this can be, for example, a threshold that is set too low or so-called clutter, a deterministic signal component caused by reflections from highly scattering objects such as rough ground or vegetation; for more information on clutter, see Sect. 1.5.3. Missed detections are targets not correctly recognized as such by the target detection algorithm. This may be due to a threshold that is set too high or a backscatter power that is too low, causing the target's signal-to-noise ratio to be too low, which means that it does not stand out sufficiently from the noise. The four possible decision types of a target detection algorithm are summarized in Table 5.5.

In contrast to a fixed threshold, a CFAR determines an adaptive local threshold individually for each cell. This adaptive threshold is estimated based on the amplitudes of surrounding cells, enabling it to reflect local signal properties such as frequency- or range-dependent behavior of the noise. A variety of CFAR algorithms can be found in the literature [2, 3]. In general, a CFAR uses a sliding window of width $N \in \mathbb{Z}$ centered around a cell under test (CUT). The width N determines the size of the neighborhood from which the threshold is to be derived. The threshold is usually calculated in two steps. In the first step, an estimator determines the ambient signal level $\bar{X}$ from the N local neighboring cells. Ideally, this value corresponds to the noise level. After that, the actual threshold is determined by multiplying $\bar{X}$ by a scaling factor $\alpha > 1$. The scaling ensures a certain safety margin against fluctuations

5.4 Target Detection using CFAR Algorithms

Table 5.5 Decision possibilities of a detection algorithm

	Target present	No target present
Target detected	Correct detection (hit)	False alarm (false positive)
No target detected	Missed detection (false negative)	Correct rejection

and variations in the noise. If $\bar{X}$ is an appropriate estimate of the expected value of the ambient noise level, then even small fluctuations would lead to false detections due to the variance in the signal if no scaling is applied. This variance is taken into account by the scaling α. Depending on the variance of the signal level, the choice of α is a compromise between the number of false detections and the risk of missing targets. Consequently, α (in dB) corresponds to the minimum SNR that a target must meet after range, velocity, and angle evaluation in order to still be detected.

The neighborhood evaluated within the sliding window around the CUT should be chosen to match the signal and target characteristics. If the neighborhood is too small, for example, if it contains multiple targets or is dominated by strong side lobes, the obtained threshold may be too high and targets may remain undetected. If the neighborhood is too large, however, not all local effects are sufficiently well represented, depending on the scenario. The sliding window is moved gradually along the radar image and a threshold is calculated for each cell. This is then compared to the value of the cell. Generally, a CFAR can be applied to each signal dimension (range, velocity or angle). Moreover, different CFAR types or different parameterizations of the algorithm can be applied in the different dimensions. Furthermore, the CFAR target detection can be performed in multiple dimensions simultaneously. The following elaborations are limited to one-dimensional CFAR algorithms, yet these can also be applied in several dimensions by combining them. The best-known CFAR algorithms are cell averaging CFAR (CA-CFAR) and ordered statistics CFAR (OS-CFAR), which are presented hereafter.

5.4.1 Cell Averaging CFAR (CA-CFAR)

The CA-CFAR determines the ambient signal level $\bar{X}$ by a simple two-level mean calculation, as illustrated in Fig. 5.9. For this purpose, the N neighboring cells around the CUT Y, $N/2$ before and after the CUT Y, are selected. Often, the cells directly adjacent to the CUT are excluded to avoid distorting the result due to correlations between neighboring cells. The excluded cells are referred to as guard cells. After that, the two mean values of the cells on either side of the CUT are computed; the mean value of which in turn results in the desired estimated value $\bar{X}^{CA}$:

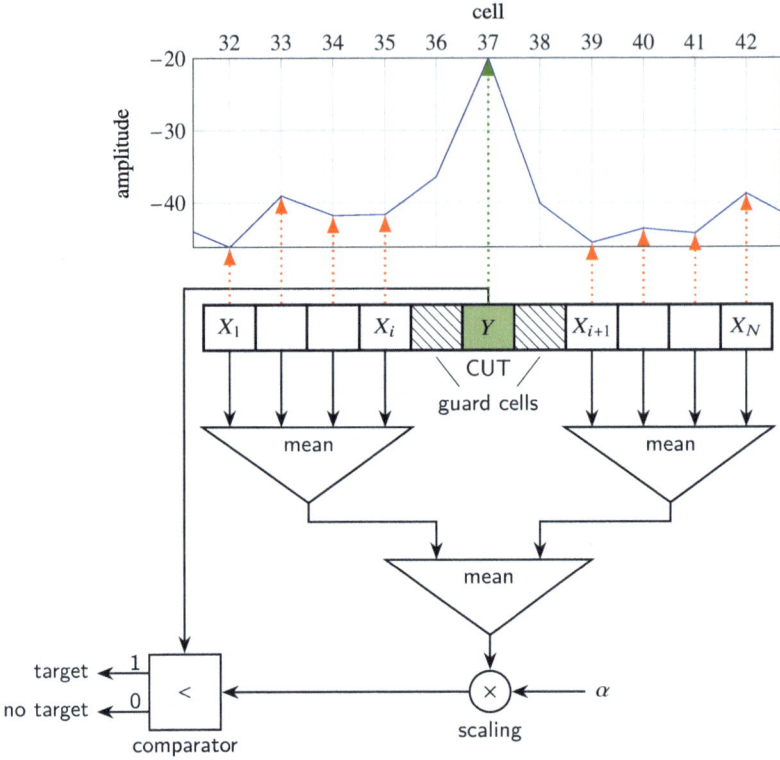

Fig. 5.9 In CA-CFAR, the cells to the right and left of the cell under test are averaged. To prevent large signal levels in the cell under test from affecting other cells, guard cells can be introduced that are not included in the average. The average is scaled by a factor α before the comparator

$$\left. \begin{array}{l} \mu_1 = \frac{1}{N/2} \sum_{i=1}^{N/2} X_i \\ \mu_2 = \frac{1}{N/2} \sum_{i=N/2+1}^{N} X_i \end{array} \right\} \Rightarrow \bar{X}^{\text{CA}} = \frac{1}{2}(\mu_1 + \mu_2) = \frac{1}{N} \sum_{i=1}^{N} X_i . \quad (5.66)$$

Finally, the estimated value $\bar{X}^{\text{CA}}$ is multiplied by the scaling factor $\alpha > 1$, which eventually yields the threshold value $\alpha \bar{X}^{\text{CA}}$ for the CUT. The CA-CFAR approach is simple and can be calculated with little effort. However, one drawback of CA-CFAR is that all amplitudes of the surrounding cells contribute equally to the threshold calculation. This makes the approach prone to errors in the presence of extended targets, since these raise the threshold considerably. A threshold that is (mistakenly) set too high leads to targets not being detected.

5.4 Target Detection using CFAR Algorithms

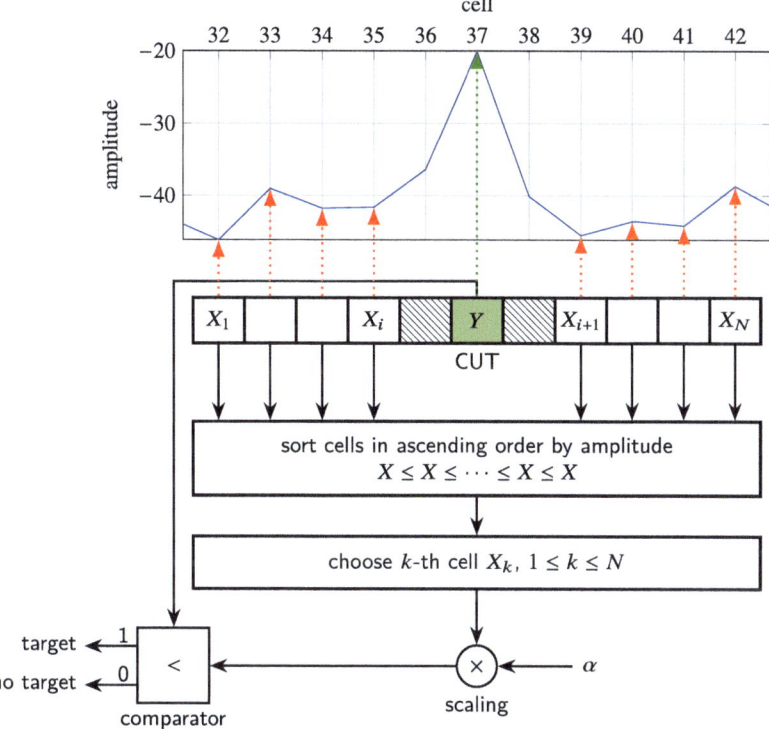

Fig. 5.10 With OS-CFAR, the cells are sorted by amplitude and then the amplitude of one cell is selected as the threshold. As with CA-CFAR, the amplitude of the cell is scaled before it is compared with the cell under test

5.4.2 Ordered Statistics CFAR (OS-CFAR)

The aforementioned problems that CA-CFAR experience can be avoided by using the OS-CFAR. The operating principle of the OS-CFAR is shown in Fig. 5.10.

In this algorithm, all N cells in the selected neighborhood—each $N/2$ before and after the CUT—are first sorted in ascending order according to their amplitude. For OS-CFAR, one or more guard cells adjacent to the CUT are also usually taken into account. In the next step, the k-th element from the sorted set is selected as the estimate $\bar{X}$:

$$\bar{X}^{OS} = X_k, \quad \text{where} \quad k = \lfloor rN \rfloor. \tag{5.67}$$

The so-called rank r, $0 \leq r \leq 1$, is adaptable, where $k = \lfloor rN \rfloor$, i.e., $r = k/N$ applies. For example, if $r = 0.5$ is chosen, then $\bar{X}$ corresponds to the median of the surrounding cells. The choice of rank influences the false alarm rate or how conservatively the algorithm behaves. For example, a rank $r < 0.5$ underestimates the (mean) signal level for Gaussian noise; On the other hand, a rank of $r > 0.5$ overestimates it. Often, a rank between 0.5 and 0.7 is chosen, since, on the one hand, the N samples often do not adequately represent the actual noise distribution. On the other hand, the values of the neighboring cells considered are only approximately Gaussian distributed or may also contain signal components of unknown distribution.

Finally, the estimated value (5.67) is also multiplied by a scaling factor $\alpha > 1$, which ultimately gives the threshold value $\alpha \bar{X}^{OS}$.

5.4.3 CFAR Target Detection

As already mentioned, in the last step of both the CA-CFAR and OS-CFAR algorithm, the estimate $\bar{X}$ is scaled by the factor $\alpha > 1$ and subsequently the amplitude of the CUT Y is compared with the scaled estimate. The binary result of this comparator is

$$\begin{cases} 1 \text{ (target)}, & \text{if } Y \geq \alpha \bar{X}^{CA|OS} \\ 0 \text{ (no target)}, & \text{if } Y < \alpha \bar{X}^{CA|OS}. \end{cases} \quad (5.68)$$

If the value of the CUT Y exceeds the threshold $\alpha \bar{X}^{CA|OS}$, it is assumed that Y is dominated by a target and not by noise. The CFAR is applied to all cells of the R-v matrix in the FoV, which eventually results in a list of potential targets.

To further refine the CFAR result, a peak detection, clustering or tracking of the targets on the list can be carried out. For example, a peak detection can eliminate peak shoulders that have exceeded the CFAR threshold. This often happens with strong targets with broadened peaks due to the window function, (see Sect. 5.2). The peaks and their shoulders then extend across several cells, with both the peak tips and part of the shoulders being detected by the CFAR. The peak detection filters out the actual peaks from the neighboring detections and thus removes shoulder detections. In contrast, clustering combines the detected targets into clusters/groups that belong together on the basis of one or more similarity measures. The aim is to group detections that originate from different objects into different clusters so that each cluster contains only reflections of one object, such as a vehicle. If this is successful, the clusters can be used, for example, for object classification. Clustering can also be used to identify possible false detections, since these usually cannot be assigned to any cluster.

References

1. H. L. van Trees, Synthesis of linear arrays and apertures, in *Optimum Array Processing*, Chap. 3. (Wiley, 2002), pp. 90–230. https://doi.org/10.1002/0471221104.ch3
2. H. Rohling, Radar CFAR thresholding in clutter and multiple target situations. IEEE Trans. Aerospace Electr. Syst. **AES-19**(4), 608–621 (1983). https://doi.org/10.1109/taes.1983.309350
3. M. Kronauge, H. Rohling, Fast two-dimensional CFAR procedure. IEEE Trans. Aerosp. Electr. Syst. **49**(3), 1817–1823 (2013). https://doi.org/10.1109/taes.2013.6558022

Open Access This chapter is licensed under the terms of the Creative Commons Attribution 4.0 International License (http://creativecommons.org/licenses/by/4.0/), which permits use, sharing, adaptation, distribution and reproduction in any medium or format, as long as you give appropriate credit to the original author(s) and the source, provide a link to the Creative Commons license and indicate if changes were made.

The images or other third party material in this chapter are included in the chapter's Creative Commons license, unless indicated otherwise in a credit line to the material. If material is not included in the chapter's Creative Commons license and your intended use is not permitted by statutory regulation or exceeds the permitted use, you will need to obtain permission directly from the copyright holder.

Chapter 6
Fundamentals of Antennas and Antenna Arrays

In many modern radar applications in the millimeter wave range, the radar must determine not only the range and velocity of a target, but also its angle at which it is located in relation to the radar. For this purpose, antenna arrays are required, i.e. a group of antennas. By evaluating and comparing the signals at different receiving antennas, the angle of incidence, i.e. the angle of the target, is determined. The larger an antenna array is in relation to the wavelength, the better its angular resolution and separation capability. For this reason, the millimeter wave range is interesting for angle estimation due to its small wavelengths. The absolute size of the sensors remains in the range of a few centimeters for most applications, even with high demands on angular resolution. While in the past, radars often employed mechanically rotating antennas with strong beam focusing to determine the angle of incidence, nowadays only electronically scanning systems are used in the millimeter wave range. Thus, a mechanical scanning system for the antenna is no longer required. The electronic approaches, known as *beamforming*, can be realized in both analog and digital domains.

After the introduction of some terms used in the context of antennas and antenna arrays in the millimeter wave range as well as the typical coordinate system in Sect. 6.1, antenna arrays are described from the point of view of an antenna engineer in Sect. 6.2. Then, a description from the point of view of signal processing follows in Sect. 6.3 in order to show the duality and differences of both approaches. In Sect. 6.4, hardware influences on array signal processing and calibration are discussed. The concept of a beamformer is presented in Sect. 6.5, which focuses the signal power in a specific direction during transmission or reception. Finally, the achievable resolution and separability of one- and two-dimensional arrays are derived in Sect. 6.6. The following Chap. 7 covers modern digital angle estimation methods as they are used nowadays.

6.1 Antenna Array Parameters and Coordinate Systems

With regard to antennas and antenna arrays, this section focuses on parameters and terms that are specifically used in the context of radar sensors in the millimeter wave range; these, as well as the typically used coordinate system for arrays, shall be presented hereafter. For general characteristic parameters to describe antennas and antenna arrays, such as radiation pattern, gain and efficiency, please refer to the standard antenna literature, such as [1].

Aperture In the context of millimeter wave radars, the maximum geometrical extent of the radiating or receiving part of the antenna array is described as the aperture. This means that signals can be received or transmitted from the aperture. The duality between transmitting and receiving applies, which means for the antennas that both are equivalently possible. For a one-dimensional array, the aperture is thus the distance between the two outermost antenna elements; for a two-dimensional array, the aperture is the area occupied by the antenna elements.

Field of View The FoV is understood to be the area in front of a radar sensor in which targets of a specified RCS can be detected with a minimum SNR. This area is described by a maximum range and two angles. The FoV can be calculated using the required application-specific minimum SNR, the radar equation for a target with the required RCS and the radiation pattern of the antennas. It should be noted that a radar sensor can also detect targets with an RCS that is greater than that assumed one outside the FoV. This often causes problems in practical applications, as it can lead to the detection of large targets outside the maximum range of the sensor. Such problems also arise in the angular domain. With an automotive radar, for example, strong reflections from the guardrails may be detected outside the FoV.

Unambiguous Angular Range The unambiguous angular range of an antenna array describes the angular range of the FoV in which targets can be unambiguously detected, i.e., no spectral repetitions occur, as explained in Sect. 3.6. However, in many applications, care must be taken to ensure that no ambiguities occur outside the FoV either, because strong targets can be mapped into the FoV by these ambiguities.

For the description of antennas or antenna arrays, spherical coordinates are usually used, as shown in Fig. 6.1. In this representation, the antenna or the antenna array is located in the origin of the coordinate system. The radial coordinate describes the distance of a point in the coordinate system to the antenna array. The coordinates ψ and ϑ describe the angles of a point in the coordinate system. In many radar applications operating in the millimeter wave spectrum, the angle ψ is typically used to describe the azimuth coordinate. This coordinate describes the angle at which a target is seen on the earth's surface, under the assumption that the radar sensor is placed on the earth's surface. The elevation angle ϑ describes the angle between the North Pole and the target, perpendicular to the azimuth angle.

6.2 Fundamental Considerations for Antenna Arrays

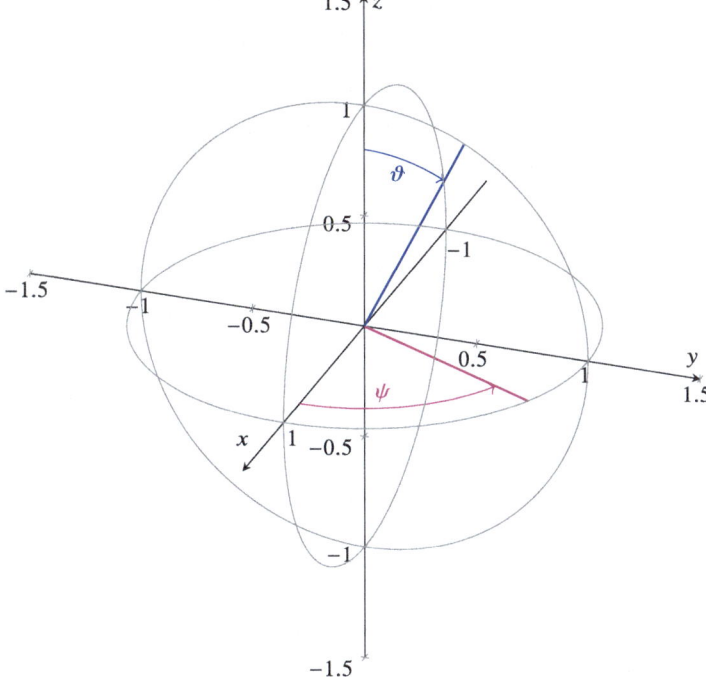

Fig. 6.1 Spherical coordinates with ψ as azimuth angle and ϑ as elevation angle

In the field of automotive radar or robotics, the azimuth angle ψ therefore describes the angle in the plane in which the vehicles move (earth's surface). This angle is important, for example, for assigning lanes to road users. The elevation angle ϑ describes how far a target is elevated from the earth's surface, where according to Fig. 6.1 $\vartheta = 0°$ corresponds to the North Pole. This angle is crucial, for example, for assessing whether an object can be passed over or under.

6.2 Fundamental Considerations for Antenna Arrays

When several antennas are combined to form an antenna array, the radiated fields of all antennas overlap when transmitting, or the incoming field is received by all antennas when receiving. Since the transmit and receive cases are dual or reciprocal, it is irrelevant which of the two cases is considered. Both cases are relevant for radars. In the following, mostly the transmit case is considered. Nevertheless, all the considerations that apply to this case can be applied to the receiving case.

If the radar is to focus its transmitted signal in different directions, the radiated fields of the individual antennas must be superimposed constructively into those directions. If the radar is to determine the angle of arrival of the signal from the received signal, the incident electromagnetic field at all antennas is analyzed with respect to the angle of arrival.

N_{ant} antennas forming the array are assumed, where m is the antenna count index. All antennas have a complex-valued feed current I_m. A point in the far field $P(R_m, \vartheta_m, \psi_m)$ is considered, which is at a range R_m from the antennas, as shown in Fig. 6.2.

The fields of the individual antennas overlap vectorially. Thus, the field strength at point P (see Fig. 6.2) is given by

$$\mathbf{E}_P = \sum_{m=1}^{N_{\text{ant}}} \mathbf{E}_m(R_m, \vartheta_m, \psi_m). \tag{6.1}$$

For the sake of simplicity, the following assumptions hold:

- All antennas have the same radiation pattern.
- All antennas are equally oriented in space, i.e. the main radiation direction is always the same.

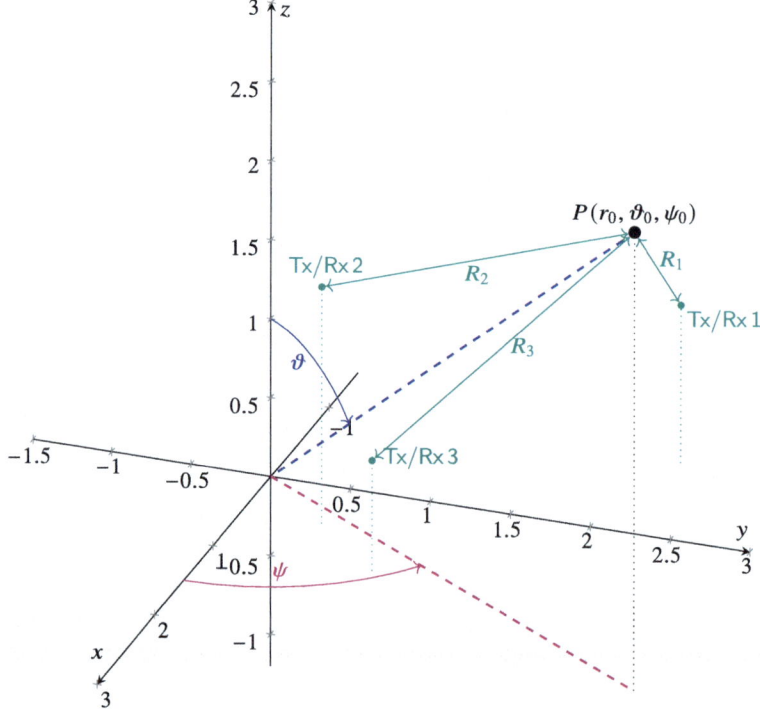

Fig. 6.2 N_{ant} antennas in the coordinate system. All fields superpose at point $P(R_0, \vartheta_0, \psi_0)$

6.2 Fundamental Considerations for Antenna Arrays

- The antennas do not influence each other, i.e. there are no coupling effects between the antennas.

Consequently, all the antennas radiate the same field and the following applies:

$$\mathbf{E}_P = \sum_{m=1}^{N_{\text{ant}}} \mathbf{E}(R_m, \vartheta_m, \psi_m, I_m). \quad (6.2)$$

The electric field of the antennas is proportional to the current I_m of the m-th antenna. Consequently, the following applies

$$\mathbf{E}_P = \sum_{m=1}^{N_{\text{ant}}} I_m \mathbf{E}'(R_m, \vartheta_m, \psi_m), \quad (6.3)$$

where $\mathbf{E}'$ is the field referenced to an arbitrarily normalized unit current. This can be further transformed into

$$\mathbf{E}_P = \sum_{m=1}^{N_{\text{ant}}} I_m \mathbf{E}''(\vartheta_m, \psi_m) \frac{e^{-j\beta_0 R_m}}{R_m}, \quad (6.4)$$

where $\mathbf{E}''$ describes a quantity that, when multiplied by $\exp\{-j\beta_0 R_m\}/R_m$, yields the original field. The normalized magnitude of $\mathbf{E}''$ thus describes the directional characteristics of the individual antennas. For very large ranges R_0 compared to the wavelength (i.e. $(R_0 - R_m) \gg \lambda$), the coordinates R_m, ϑ_m, and ψ_m can be replaced by the corresponding quantities R_0, ϑ_0, and ψ_0 of the reference coordinate system for the amplitude terms. However, this does not apply to the phase term $\exp\{-j\beta_0 R_m\}$. The exponential function with an imaginary argument is periodic, so even small differences in R_m or antenna positions lead to relevant phase changes. The electric field (6.4) for large R_0 is thus approximated by:

$$\mathbf{E}_P = \sum_{m=1}^{N_{\text{ant}}} I_m \mathbf{E}''(\vartheta_0, \psi_0) \frac{e^{-j\beta_0 R_m}}{R_0} = \mathbf{E}''(\vartheta_0, \psi_0) \sum_{m=1}^{N_{\text{ant}}} I_m \frac{e^{-j\beta_0 R_m}}{R_0}. \quad (6.5)$$

After introducing a reference current I_0, (6.5) can be furthermore rewritten as

$$\mathbf{E}_P = \underbrace{\mathbf{E}''(\vartheta_0, \psi_0)}_{F_E} \underbrace{I_0 \frac{e^{-j\beta_0 R_0}}{R_0}}_{F_A} \underbrace{\sum_{m=1}^{N_{\text{ant}}} \frac{I_m}{I_0} e^{-j\beta_0 (R_m - R_0)}}_{F_G}. \quad (6.6)$$

This expression describes the field strength at point P in the far field of the antenna array and consists of three terms:

- The *element factor* $\mathbf{F}_E = \mathbf{E}''(\vartheta_0, \psi_0) I_0$ is determined by the directional characteristics of the antennas, i.e. the radiation pattern.
- The *distance factor* $F_A = \exp\{-j\beta_0 R_0\}/R_0$ describes the wave propagation in free space.
- The *group factor*

$$F_G = \sum_{m=1}^{N_{\text{ant}}} \frac{I_m}{I_0} e^{-j\beta_0 (R_m - R_0)} \qquad (6.7)$$

describes the placement of the antennas within the array and the feeding of the individual antennas. However, it is independent of the type of antenna or its directional characteristics. Ultimately, it describes the radiation of the array if it would consist of isotropic radiators.

The radiation pattern of the antenna array is therefore the product of the group factor and the pattern of the individual antenna elements, normalized to one. For further considerations, it is often sufficient to consider only the group factor, since it is the only one that is influenced by the array arrangement and the feeding of the antennas.

The mathematical interpretation of the group factor is crucial for understanding the operating principle of antenna arrays. The group factor (6.7) corresponds to a discrete Fourier transform of the discrete current distribution function, assuming the values I_m/I_0 at the sampling points. I_m/I_0 is often referred as to antenna weight or antenna coefficient in this context. An antenna array can thus be interpreted as the spatial sampling of a current distribution function, whose Fourier transform represents the normalized far field. This in turn means that a very large array leads to a strong focusing, and conversely, a small array tends to radiate broadly. Since each antenna in the array is individually controlled via I_m, practically any current distribution can be discretely represented and thus a large variety of radiation patterns can be generated.

In the receive case, the Fourier relationship implies that the discrete current distribution function and the incident field form a Fourier pair. Consequently, the Fourier transform of the current distribution function can be used to deduce the incident field and thus the angle of arrival, see Sect. 7.2. For the transmit case, this means that the Fourier relationship can be used to calculate which discrete current distribution function or antenna coefficients must be set to achieve a specific far field.

Consequently, the design of an antenna array needs to take into account the sampling theorem in order to avoid ambiguities, so-called grating lobes. This is achieved by selecting an antenna spacing d with

$$d \leq \frac{\lambda}{2}. \qquad (6.8)$$

6.3 Description of Antenna Arrays Using a Signal Model

The electric field components and currents on the antennas introduced Sect. 6.2 can be interpreted as input and output signals for an antenna array, depending on the transmit or receive case. The conversion of field into currents or vice versa depends on the antenna impedances and the load or source impedances connected to the antennas. Further details can be found in [2]. In this section, a signal model for an antenna array is developed that allows the description of the array solely based on input and output signals. The antennas are assumed to be ideal in every respect, i.e. power matching, the narrowband approximation and all previous assumptions from Sect. 6.2 hold. For the sake of clarity, the following explanations are limited to one-dimensional arrays.

It is assumed that N_{ant} antennas of the array are arranged along the x-axis of the coordinate system, as shown in Fig. 6.3. The azimuth angle ψ describes the direction of arrival of a plane wave. At $\psi = 90°$ the wave impinges perpendicularly to the array. The signal $g(t)$ describes the incident plane wave reflected from a target while satisfying the far-field condition. This results in the received signal at the m-th receive antenna of

$$y_m(t) = a_m(\psi)g(t) + n_0(t), \quad m \in [1; N_{ant}] . \tag{6.9}$$

$a_m(\psi)$ is an angular weighting factor that describes the behavior of the m-th antenna with its respective radiation pattern, as well as the behavior of the receiver; $n_0(t)$ describes the additive white noise of the receiver.

For an antenna array with N_{ant} antennas, this can be written as a vector equation, where the elements of the vector describe the signals at the individual antennas

$$\mathbf{y}(t) = \mathbf{a}(\psi)g(t) + \mathbf{n_0}(t) . \tag{6.10}$$

The vector $\mathbf{a}$ is known by a variety of names in the literature; this book uses the commonly used name *steering vector*. The steering vector $\mathbf{a}$ describes the amplitudes and phases that result at the individual antennas when the plane wave impinges. Figure 6.3 shows that the plane wave travels different distances to the individual antennas. The difference, relative to an antenna at the origin of the coordinate system, is called the path difference and results from

$$G = d_{m,x} \cos(\psi), \tag{6.11}$$

where $d_{m,x}$ describes the position of the m-th antenna along the x-axis. This can be used to determine the phase at the m-th antenna relative to an antenna at the origin of the coordinate system with

$$\phi_m = \frac{2\pi}{\lambda_0} G = \frac{2\pi}{\lambda_0} d_{m,x} \cos(\psi) . \tag{6.12}$$

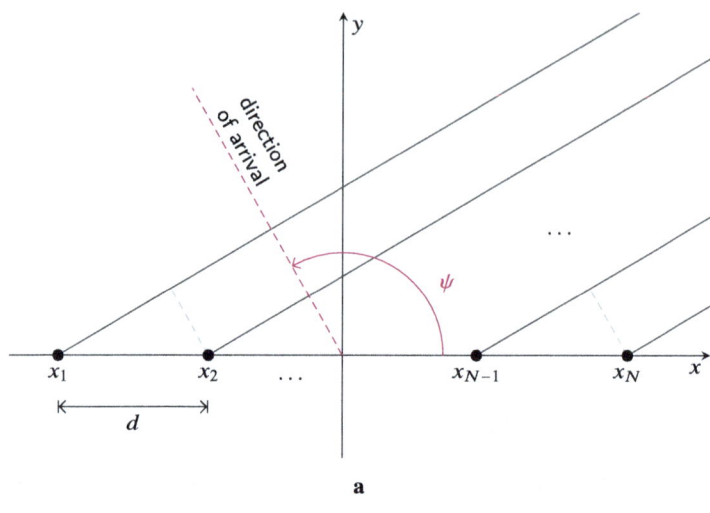

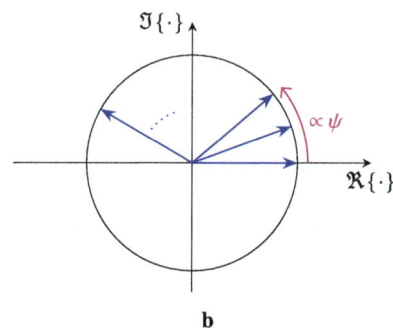

Fig. 6.3 ULA: **a** schematic of a ULA with antenna spacing d and an exemplary impinging wave with angle of incidence ψ (*magenta*). The *black solid lines* resemble the wave fronts at the antenna positions and the *gray dashed lines* represent the path difference between adjacent antennas. **b** Exemplary corresponding phasors (*blue*) of the waves detected at the antennas, normalized to the antenna at x_1. The phase progression along the antennas is proportional to the angle ψ

This means that the steering vector **a** can be rewritten as

$$\mathbf{a}(\psi) = \left[A_1(\psi) e^{j\frac{2\pi}{\lambda_0} d_{1,x} \cos(\psi)}, \ldots, A_{N_{\text{ant}}}(\psi) e^{j\frac{2\pi}{\lambda_0} d_{N_{\text{ant}},x} \cos(\psi)} \right]. \quad (6.13)$$

A_m denotes the amplitude of the received signal at the m-th antenna. If there are multiple targets in the channel, multiple plane waves impinge onto the antenna array, all coherently superimposed. This yields for Q targets

6.5 Beamformer and Power Evaluation

$$\mathbf{y}(t) = \sum_{q=1}^{Q} \mathbf{a}(\psi_q) g_q(t) + \mathbf{n_0}(t). \quad (6.14)$$

6.4 Calibration and Hardware Influences

So far, the antennas and receivers have been assumed to be ideal, i.e. all having the same behavior. However, especially in the millimeter wave range, each receive channel shows a slightly different behavior in practice, due to the small wavelength and large tolerances of the hardware. In particular, due to tolerances, the electrical path length to the individual antennas is often different, even if the design provides for identical electrical lengths. Furthermore, the coupling between the individual antennas, which usually occurs in practice, leads to deterministic, often angle-dependent phase and amplitude errors, which must be compensated for an accurate angle estimation. Calibration of the antenna arrays and receivers is required for this reason. However, these errors are not taken into account in the signal model in (6.13). To that end, a more general modeling of the steering vector like

$$\mathbf{a}^T(\psi) = \begin{bmatrix} C_1(\psi) A_1(\psi) e^{j\frac{2\pi}{\lambda_0} d_{1,x} \cos(\psi) + \Xi_1(\psi)} \\ \vdots \\ C_N(\psi) A_{N_{\text{ant}}}(\psi) e^{j\frac{2\pi}{\lambda_0} d_{N_{\text{ant}},x} \cos(\psi) + \Xi_{N_{\text{ant}}}(\psi)} \end{bmatrix} \quad (6.15)$$

must be applied, where $\Xi_m(\psi)$ and $C_m(\psi)$ describe the amplitude and phase errors of the m-th antenna. In general, both quantities are a function of the angle ψ, but the amplitude term can often be ignored. During calibration, the phase error $\Xi_m(\psi)$ and, if necessary, the amplitude error $C_m(\psi)$ are determined so that they can be compensated for to allow further signal processing. In the calibration measurement, a strong target is usually positioned in an anechoic chamber. The error terms are determined as a function of the azimuth angle ψ by an angle-dependent measurement of the target with the radar sensor. If there are no systematic errors over m, averaging of the errors occurs in large arrays with a large number of elements, so that angle-dependent calibration becomes less important with an increasing number of channels. If the angular dependency can be neglected, a calibration using a single stationary target is sufficient.

6.5 Beamformer and Power Evaluation

So-called *beamformers* are used to focus signals in a specific direction at the transmitter or to determine the direction of arrival at the receiver. On the transmit side, the transmitted signals of the individual antennas are controlled in such a way that the fields constructively superpose in a certain direction in the far field, i.e. the main beam direction of the antenna array is steered in a certain direction. On the receive side, for example, it is analyzed from which direction which amount of power is received. The targets can be found at the angles with a power maximum.

For this purpose, different algorithms are used in the digital domain, i.e. after the analog-digital conversion of the signals. These algorithms are discussed in Chap. 7. Here, a general mathematical model of a beamformer is presented, supplemented by the resulting calculation of the received power.

Mathematically, a beamformer can be described by

$$s(t) = \sum_{m=1}^{N_{\text{ant}}} w_m^*(\psi) y_m(t) = \mathbf{w}^\dagger(\psi) \cdot \mathbf{y}(t), \quad m \in [1; N_{\text{ant}}] \ . \tag{6.16}$$

In this case, w_m describes the coefficient by which the signal $y_m(t)$ of the m-th antenna is weighted in amplitude and phase before all N_{ant} signals $y_m(t)$ are added up. This results in the output signal $s(t)$ of the beamformer in the receive case. The product $w_m^* y_m(t)$ corresponds to the antenna coefficients of the current distribution function in (6.7). In the transmit case, the vector $\mathbf{w}$ contains the antenna coefficients, if the same transmit signals are distributed to all antennas.

The output power of the beamformer is calculated from the output signal $s(t)$ as

$$P(\mathbf{w}) = \mathrm{E}\{|s(t)|\}^2 \ . \tag{6.17}$$

$\mathrm{E}\{\cdot\}$ represents the expectation operator. In practice, the expected value is approximated by considering the past M time steps. This results in the output power

$$\begin{aligned} P(\mathbf{w}) &= \mathbf{w}^\dagger \cdot \mathrm{E}\{\mathbf{y}(t) \cdot \mathbf{y}^\dagger(t)\} \cdot \mathbf{w} \\ &\approx \frac{1}{M} \sum_{m=1}^{M} |s_m(t)|^2 = \frac{1}{M} \sum_{m=1}^{M} \mathbf{w}^\dagger \cdot \mathbf{y}_m(t) \cdot \mathbf{y}_m^\dagger(t) \cdot \mathbf{w}, \end{aligned} \tag{6.18}$$

by averaging over M time steps. The covariance matrix of the received signal $\mathbf{y}(t)$ is

$$\mathbf{R} = \mathrm{E}\{\mathbf{y}(t) \cdot \mathbf{y}^\dagger(t)\}, \tag{6.19}$$

which in turn is approximated by considering the last M time steps:

$$\mathbf{R} \approx \frac{1}{M} \sum_{m=1}^{M} \mathbf{y}_m(t) \cdot \mathbf{y}_m^\dagger(t) \ . \tag{6.20}$$

6.6 Angular Resolution and Angular Separability

Similar to the resolution of a radar in the measurement dimensions of velocity and range, as explained in the Chap. 3, the resolution limit for an antenna array can be derived from the Fourier relationship between the current distribution and the far field. To this end, the angular distance between the two zeros to the left and right of the signal maximum in the direction of the angle of arrival or the main beam direction is used. This distance is referred to as the Rayleigh criterion and is a measure of the resolution of an array. Alternatively, the resolution of an antenna array can be characterized by the angular resolution according to Sect. 3.3.3 with

$$|C(\psi)|^2 = 0.5. \tag{6.21}$$

Analogous to Sect. 3.4, the angular separation capability is obtained from twice or three times the value, depending on the application context. All these quantities must be calculated for each antenna arrangement in the array and main beam direction from (6.6). In practice, however, linear, i.e. one-dimensional, or planar, i.e. two-dimensional, antenna arrays with equidistant antennas are used frequently. They are referred to as uniform linear arrays (ULAs) and uniform rectangular arrays (URAs). For these special cases, the resolution can be calculated in a closed form, as shown below.

6.6.1 One-Dimensional Antenna Arrays

For a ULA consisting of isotropic antennas with an antenna spacing d as shown in Fig. 6.3, the radiation pattern can be determined from (6.6) with

$$C(\psi) = \frac{1}{N_{\text{ant}}} \frac{\sin\left(N_{\text{ant}} \frac{\pi d}{\lambda}(\cos\psi - \cos\psi_0)\right)}{\sin\left(\frac{\pi d}{\lambda}(\cos\psi - \cos\psi_0)\right)}, \quad 0 \leq \psi \leq \pi, \tag{6.22}$$

where ψ_0 represents the main beam direction.

For the sake of simplicity and to eliminate the dependence of almost all array characteristics on $\cos(\psi)$, a coordinate u is usually introduced where $u = \cos(\psi)$. This simplifies (6.22) to

$$C(\psi) = \frac{1}{N_{\text{ant}}} \frac{\sin\left(N_{\text{ant}} \frac{\pi d}{\lambda}(u - u_0)\right)}{\sin\left(\frac{\pi d}{\lambda}(u - u_0)\right)}, \quad -1 \leq u \leq 1. \tag{6.23}$$

The angular resolution according to (6.21) can be approximated for $N_{\text{ant}} > 10$, assuming the main beam direction is perpendicular to the array, by

$$\Delta u = u - u_0 = 0.891 \frac{\lambda}{N_{\text{ant}} d}. \tag{6.24}$$

The larger N_{ant} gets, the smaller the coefficient in (6.24) becomes. For $N_{\text{ant}} > 30$, it is approximately 0.886. For an angle $\psi = \psi_0 + \Delta\psi$, which is given by ψ_0 plus the minimum increment $\Delta\psi$, follows from (6.24) that

$$\Delta u = \cos(\psi_0 + \Delta\psi) - \cos(\psi_0) = 0.891 \frac{\lambda}{N_{\text{ant}} d}. \tag{6.25}$$

For the main beam direction $\psi_0 = \pi/2$, the angular resolution is

$$\Delta\psi_{3\text{dB}} = \arccos\left(0.891 \frac{\lambda}{N_{\text{ant}} d}\right) - \frac{\pi}{2}. \tag{6.26}$$

In the context of automotive radar applications in particular, the angle of arrival is sometimes defined in the azimuth plane relative to the perpendicular incidence on the array. In this case, $\tilde{\psi} = \psi + \pi/2$ and $\tilde{\psi}_0 = 0$ and thus

$$\Delta\tilde{\psi}_{3\text{dB}} = \arcsin\left(0.891 \frac{\lambda}{N_{\text{ant}} d}\right). \tag{6.27}$$

The zeros in (6.23) result from

$$\sin\left(\frac{\pi N_{\text{ant}} d}{\lambda} u\right) = 0 \tag{6.28}$$

respectively

$$\frac{\pi N_{\text{ant}} d}{\lambda} u = m\pi, \quad m = 1, 2, \ldots \tag{6.29}$$

This means that the distance of the first two zeros to the left and right of the maximum of the radiation pattern is

$$\Delta u_R = \frac{\lambda}{N_{\text{ant}} d} \tag{6.30}$$

as the Rayleigh criterion and thus

$$\Delta\psi_{\text{ray}} = \arccos \frac{\lambda}{N_{\text{ant}} d} - \frac{\pi}{2}. \tag{6.31}$$

In practice, both the Rayleigh criterion and the resolution definition according to (6.21) and (6.26) are referred to as angular resolution. The resolution improves when the antenna spacing d and thus the array size or aperture increases. This again shows the Fourier relationship between current distribution and far field. It should also be noted that the resolution is highest when measured perpendicular to the array orientation, since (6.26) only applies to a main beam direction perpendicular to the array. Otherwise, the angular resolution decreases with $\cos \psi$.

However, increasing the distance d between the antennas to improve the resolution has its limits. Equation (6.22) is only unambiguous if

$$\frac{\pi d}{\lambda} \leq \frac{\pi}{2} \tag{6.32}$$

holds. This means that

$$d \leq \frac{\lambda}{2}, \tag{6.33}$$

i.e. the antenna spacings must be less than or equal to half a wavelength. As already mentioned in Sect. 6.2, this corresponds to a compliance with the sampling theorem. However, the unambiguity of the pattern is already given if (6.33) is fulfilled once in an array with non-equidistant antenna spacings; the other antenna spacings may be greater. However, antenna spacings that are not equidistant lead to higher side lobes, which result in a PSLR of less than 13 dB, which is achieved by a rectangular current distribution function. The design of antenna arrays is discussed in more detail in Sect. 8.4.

6.6.2 Two-Dimensional Antenna Arrays

In the two-dimensional case, only URAs are considered here. For a more specific treatment of such two-dimensional arrays, please refer to related literature [3]. In the following, it is assumed that the aperture extends over the x-z plane in the coordinate system as shown in Fig. 6.4, which results in

$$\mathbf{u}^\mathsf{T} = \begin{bmatrix} \sin(\vartheta)\cos(\psi) \\ \sin(\vartheta)\sin(\psi) \\ \cos(\vartheta) \end{bmatrix}. \tag{6.34}$$

The radiation pattern of an isotropic array with $N_{\text{ant},x}$ antennas of spacings d_x in the x direction, and $N_{\text{ant},z}$ antennas of spacings d_z in the z direction is

$$C(\psi, \vartheta) = \frac{\sin(N_{\text{ant},x} \pi \frac{d_x}{\lambda}(u_x - u_{x0}))}{N_{\text{ant},x} \cdot \sin(\pi \frac{d_x}{\lambda}(u_x - u_{x0}))} \cdot \frac{\sin(N_{\text{ant},z} \pi \frac{d_z}{\lambda}(u_z - u_{z0}))}{N_{\text{ant},z} \cdot \sin(\pi \frac{d_z}{\lambda}(u_z - u_{z0}))}. \tag{6.35}$$

Here, u_x and u_z correspond to the vector components from (6.34) and u_{x0} and u_{z0} represent the main beam directions (ψ_0, ϑ_0). The resolution according to (6.21) transferred to the two-dimensional case cannot be given in a closed form, as in

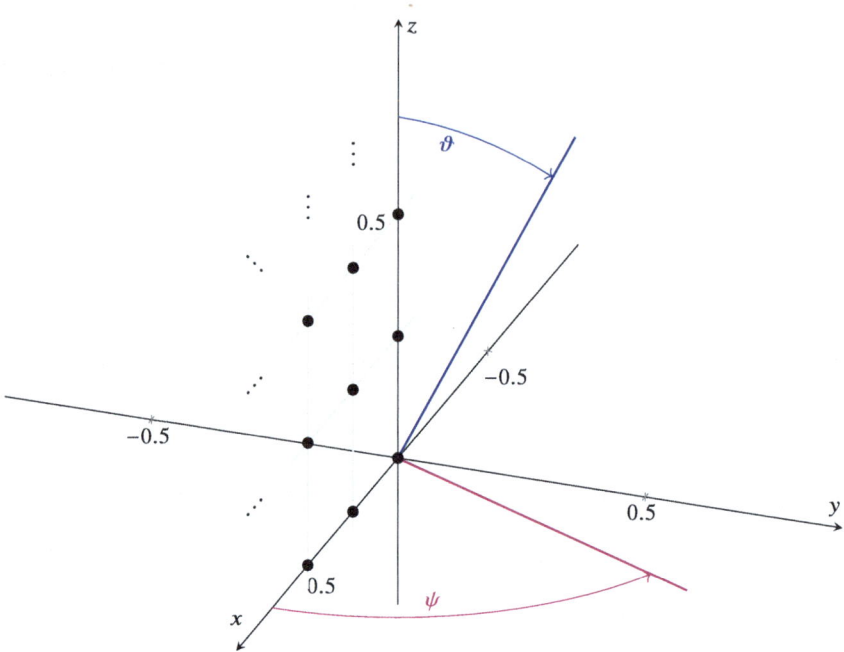

Fig. 6.4 Uniform rectangular array (URA) with antennas in the x-z plane

the one-dimensional case. However, the resolution for a given main beam direction (ψ_0, ϑ_0) can be approximated by

$$\Delta\psi_{3\text{dB}} = \cos^{-1}\left(\frac{u_{x0} - 0.443\frac{\lambda}{N_{\text{ant},x}d_x}}{\sin(\vartheta_0)}\right) - \cos^{-1}\left(\frac{u_{x0} + 0.443\frac{\lambda}{N_{\text{ant},x}d_x}}{\sin(\vartheta_0)}\right) \quad (6.36)$$

and

$$\Delta\vartheta_{3\text{dB}} = \cos^{-1}\left(u_{z0} - 0.443\frac{\lambda}{N_{\text{ant},z}d_z}\right) - \cos^{-1}\left(u_{z0} + 0.443\frac{\lambda}{N_{\text{ant},z}d_z}\right). \quad (6.37)$$

In the context of angle estimation, which is discussed in Chap. 7, the radiation pattern is not a familiar term; instead, one refers to the angular spectrum, which then replaces the radiation pattern in (6.35) and (6.22). From (6.35), the unambiguous angular range of the two-dimensional uniform array follows with

$$|u_x| \leq \frac{\lambda}{2d_x} \quad \text{and} \quad |u_z| \leq \frac{\lambda}{2d_z}. \quad (6.38)$$

References

1. C. Balanis, *Modern Antenna Handbook*. (Wiley, 2007). ISBN: 9780470294154. https://doi.org/10.1002/9780470294154
2. N. Geng, W. Wiesbeck, *Planungsmethoden für die Mobilkommunikation* (Springer, Berlin Heidelberg, 1998). ISBN: 9783642589805. https://doi.org/10.1007/978-3-642-58980-5
3. H.L. Van, *Trees, Optimum Array Processing*. (Wiley, 2002). ISBN: 9780471221104. https://doi.org/10.1002/0471221104

Open Access This chapter is licensed under the terms of the Creative Commons Attribution 4.0 International License (http://creativecommons.org/licenses/by/4.0/), which permits use, sharing, adaptation, distribution and reproduction in any medium or format, as long as you give appropriate credit to the original author(s) and the source, provide a link to the Creative Commons license and indicate if changes were made.

The images or other third party material in this chapter are included in the chapter's Creative Commons license, unless indicated otherwise in a credit line to the material. If material is not included in the chapter's Creative Commons license and your intended use is not permitted by statutory regulation or exceeds the permitted use, you will need to obtain permission directly from the copyright holder.

Chapter 7
Methods for Angle Estimation

Nowadays, only digital angle estimation methods are used in modern radar sensors in the millimeter wave range. The most important approaches to digital angle estimation are therefore presented in this chapter. First, in Sects. 7.1 and 7.2, the two basic approaches to angle determination by correlation or alternatively by a Fourier transform are discussed. These two approaches have an high practical relevance, as they are used very often. These approaches are often combined with more powerful but also more complex approaches, as presented in Sect. 7.3, to achieve high-resolution and at the same time robust, i.e. reliable, angle estimation.

7.1 Angle Estimation by Correlation

When determining the angle by correlation, no physical antenna model is used. The basic idea is to compare or correlate the measured steering vector $\mathbf{a}(\psi)$ with reference steering vectors for each angle of arrival. The physical characteristics of the antenna array are contained in the reference steering vectors as a function of angle. These reference vectors are recorded in advance in a calibration measurement (see Sect. 6.4) and stored in a calibration matrix $\mathbf{A}_{\text{cal}}(\psi)$. The angular spectrum $P(\psi)$, in which the target angles can be recognized as peaks, is given by the square of the cross-correlation of the measured steering vector with the unknown angle of arrival and the calibration matrix, i.e.

$$s_{\text{cor}}(\psi) = \frac{|\mathbf{A}_{\text{cal}}(\psi) \cdot \mathbf{a}^{\dagger}(\psi)|}{\|\mathbf{A}_{\text{cal}}(\psi)\| \, \| \mathbf{a}(\psi) \|} \tag{7.1}$$

and

$$P_{\text{cor}}(\psi) = |s_{\text{cor}}(\psi)|^2, \tag{7.2}$$

© The Author(s) 2025
C. Waldschmidt et al., *Millimeter Wave Radar*,
https://doi.org/10.1007/978-3-031-89118-2_7

where

$$\mathbf{A}_{\text{cal}}(\psi) = \begin{bmatrix} \mathbf{a}_{\text{cal}}(\psi = \psi_{\text{start}}) \\ \vdots \\ \mathbf{a}_{\text{cal}}(\psi = \pi/2) \\ \vdots \\ \mathbf{a}_{\text{cal}}(\psi = \psi_{\text{stop}}) \end{bmatrix}. \qquad (7.3)$$

A high correlation value indicates a high degree of similarity between the measured steering vector and the reference steering vector from the calibration matrix, whereas a low correlation value indicates a low degree of similarity. When using this simple but very robust method, it must be considered that the correlation can only be determined for angles for which a reference steering vector is available from a calibration measurement. To exploit the resolution capabilities of the antenna array, the angular step size of the calibration matrix should be significantly smaller than the angular resolution of the antenna array.

In practice, angle estimation by correlation has proven to be a very robust method in the millimeter wave range, since typical error influences due to the frequency range, such as manufacturing tolerances or antenna couplings, are implicitly taken into account in the calibration matrix $\mathbf{A}_{\text{cal}}(\psi)$, thus limiting their disturbing influence. The disadvantage of this approach is the high measurement effort required to create the calibration matrix. This is particularly relevant when large antenna arrays with good angular resolution are used, since a particularly large number of measurements are required to create the calibration matrix in this case.

7.2 Angle Estimation by Fourier Transform

The aim of angle estimation based on FT is to combine all signals in-phase for a certain angle and thus obtain the power distribution as a function of angle. To do this, all signal delays, i.e. phase shifts, must be compensated for before the signals are accumulated. In the case of a one-dimensional antenna array with equal antenna spacings, i.e. a ULA, arranged along the x-axis as in Fig. 6.3, the calculation of the angle-dependent power distribution can be efficiently carried out by an DFT. In the case of a ULA, the phase of the received signal changes linearly from antenna to antenna depending on the angle of incidence ψ, according to (6.13). The phase values can thus be multiplied with a complex weighting factor that increases linearly. This results in a constructive superposition in the case of the correctly assumed angle of arrival. The angular spectrum is calculated at the ξ-th discrete sampling point in the angle spectrum to

7.3 Angle Estimation Algorithms

$$s_{\xi,\text{FFT}}(\xi) = \sum_{m=0}^{N_{\text{ant}}-1} a_m e^{-j2\pi \frac{\xi}{N_{\text{ant}}} m} \quad (7.4)$$

and

$$P_{\xi,\text{FFT}} = |s_{\xi,\text{FFT}}|^2. \quad (7.5)$$

The number of samples in the angular spectrum results from the length of the DFT, which is N_{ant} for a ULA. Often, the DFT is calculated with zero padding, resulting in a smoothed angle spectrum. Howerver, zeropadding does not affect the angular resolution of the beamformer. From (6.13) it becomes clear that the phase progression, i.e. the change of the phase across the antenna array, is only linear as a function of $\cos(\psi)$. Thus, the angular spectrum has a cosinusoidal dependency, so that the frequency vector associated with the angular spectrum has to be scaled by

$$\psi_\xi \in \arccos\left[-\frac{1}{2}\frac{\lambda}{d_x}, \frac{1}{2}\frac{\lambda}{d_x}\right], \quad (7.6)$$

where d_x is the constant spacing between adjacent antennas along the x-axis.
If no ULA is used or if there are deviations from the ideal antenna positions due to manufacturing tolerances, the realization of the FT in (7.4) must be adapted accordingly to irregular antenna spacings.

The advantages of the FT-based beamformer are its simple implementation, low computational complexity and very low calibration effort, since, in contrast to the correlation-based angular estimation in Sect. 7.1, no angle-dependent calibration matrix has to be recorded. A disadvantage is its limited angle resolution, which is limited by the aperture, analogous to angle estimation by correlation.

7.3 Angle Estimation Algorithms

The following angle estimation algorithms for a digital beamforming system are the best-known representatives from different complexity and performance classes and are representative of a multitude of similar algorithms. In particular, in the field of high-resolution angle estimation methods, a large number of algorithms have been developed in the last two decades, which can be found in the relevant specialized literature.

7.3.1 Bartlett Beamformer

The Bartlett beamformer calculates the steering vector **w** in (6.16) such that the average output power of the beamformer is maximal. In contrast to the estimation with a FT, as shown in Sect. 7.2, the Bartlett beamformer takes into account M time steps or measurements, i.e. it performs a power determination according to (6.18). The received power P as an average over M time steps is given by

$$P(\mathbf{w}) = \frac{1}{M}\sum_{m=1}^{M}|s_m(t)|^2 = \frac{1}{M}\sum_{m=1}^{M}\mathbf{w}^\dagger \mathbf{y}_m(t) \cdot \mathbf{y}_m^{\dagger}(t)\mathbf{w} \qquad (7.7)$$

where, as in Sect. 6.5, it is assumed that

$$\frac{1}{M}\sum_{m=1}^{M}\mathbf{w}^\dagger \mathbf{y}_m(t) \cdot \mathbf{y}_m^{\dagger}(t)\mathbf{w} \approx E\{\mathbf{w}^\dagger \cdot \mathbf{y}(t) \cdot \mathbf{y}^\dagger(t) \cdot \mathbf{w}\} = \mathbf{w}^\dagger \cdot \mathbf{R} \cdot \mathbf{w} \qquad (7.8)$$

applies, where **R** is the covariance matrix (6.19) of the received signal $\mathbf{y}(t)$.

The output power of the beamformer should now be maximized, i.e.,

$$\max_{w}\{E\{\mathbf{w}^\dagger \cdot \mathbf{y}(t) \cdot \mathbf{y}^\dagger(t) \cdot \mathbf{w}\}\} = \max_{w}\{\mathbf{w}^\dagger E\{\mathbf{y}(t) \cdot \mathbf{y}^\dagger(t)\}\mathbf{w}\} \qquad (7.9)$$

$$= \max_{w}\{E\{|s(t)|^2\}|\mathbf{w}^\dagger \cdot \mathbf{a}(\psi)|^2 + \sigma^2|\mathbf{w}|^2\}, \qquad (7.10)$$

where the noise covariance matrix equals

$$E\{\mathbf{n_0}(t) \cdot \mathbf{n_0}^\dagger(t)\} = \sigma^2 \mathbf{I}. \qquad (7.11)$$

Here, **I** denotes the unit matrix, i.e. it is assumed that the noise is uncorrelated in all channels.

To avoid the trivial solution, the maximization of (7.10) must be subject to a boundary condition. Therefore, without loss of generality, $\|\mathbf{w}\|=1$ is defined. This means that the beamformer does not change the received power. It is easy to show that the maximum power in (7.10) subject to the boundary condition $\|\mathbf{w}\|=1$ is given by

$$\mathbf{w}_{\text{Bartlett}} = \frac{\mathbf{a}(\psi)}{\sqrt{\mathbf{a}^\dagger(\psi) \cdot \mathbf{a}(\psi)}}. \qquad (7.12)$$

This steering vector $\mathbf{w}_{\text{Bartlett}}$ can be interpreted as an angular MF that is adapted to the angle of arrival of the signal, as in the estimation via a FT (see Sect. 7.2). Finally, the individual signal delays at the antennas are compensated so that all signals are added in-phase and thus are constructively superimposed.

7.3 Angle Estimation Algorithms

Substituting (7.12) into (6.18) yields the angular spectrum

$$P_{\text{Bartlett}}(\psi) = \frac{\mathbf{a}^\dagger(\psi) \cdot \mathbf{R} \cdot \mathbf{a}(\psi)}{\mathbf{a}^\dagger(\psi) \cdot \mathbf{a}(\psi)}. \qquad (7.13)$$

In the case that only one measurement was performed ($M = 1$), the calculations via the FT according to Sect. 7.2 and the Bartlett beamformer yield identical results.

7.3.2 Capon Beamformer

The Capon beamformer reverses the idea of the Bartlett beamformer by keeping the signal power at the output of the beamformer constant for a given angle of arrival (boundary condition: $\mathbf{w}^\dagger \cdot \mathbf{a}(\psi) = 1$), while checking whether the total power from all directions is minimal. Thus, the total power in the angular spectrum is minimized ($\min\{P(\mathbf{w})\}$) under the just described constraint. In mathematical terms, therefore, the following applies:

$$\min\{P(\mathbf{w})\} \quad \text{with} \quad \mathbf{w}^\dagger \cdot \mathbf{a}(\psi) = 1, \qquad (7.14)$$

where $P(\mathbf{w})$ results from (6.18). The minimization problem in (7.14) can be solved, for example, by the Lagrangian method. This results in the steering vector

$$\mathbf{w}_{\text{Capon}} = \frac{\mathbf{R}^{-1} \cdot \mathbf{a}(\psi)}{\mathbf{a}^\dagger(\psi) \cdot \mathbf{R}^{-1} \cdot \mathbf{a}(\psi)}. \qquad (7.15)$$

If (7.15) is used in (6.18), the angular spectrum results in

$$P_{\text{Capon}}(\psi) = \frac{1}{\mathbf{a}^\dagger(\psi) \cdot \mathbf{R}^{-1} \cdot \mathbf{a}(\psi)}. \qquad (7.16)$$

The advantage of the Capon beamformer lies in the significantly higher resolution compared to the Bartlett beamformer and the FT. However, the Capon beamformer might become unstable due to the necessary inversion of the covariance matrix.

7.3.3 High-Resolution Methods: MUSIC

MUltiple SIgnal Classification (MUSIC) is presented here as representative of a high-resolution angle estimation method or beamformer. MUSIC differs fundamentally from the approaches presented so far in this section. With MUSIC, the angular resolution no longer depends directly on the aperture size, thus achieving significantly better angular separability. However, the computational effort for MUSIC is many times higher than for the previously presented approaches.

To derive the MUSIC algorithm, the signal model from (6.14) must first be converted to matrix notation:

$$\mathbf{y}(t) = \mathbf{A}(\psi) \cdot \mathbf{g}(t) + \mathbf{n_0}(t), \tag{7.17}$$

where

$$\mathbf{A}(\psi) = \begin{bmatrix} \mathbf{a}(\psi_1), \ldots, \mathbf{a}(\psi_Q) \end{bmatrix}, \tag{7.18}$$

$$\mathbf{g}(t) = \begin{bmatrix} g_1(t), \ldots, g_Q(t) \end{bmatrix}^\top. \tag{7.19}$$

The covariance matrix of the received signals is estimated by

$$\mathbf{R} = E\{\mathbf{y}(t) \cdot \mathbf{y}^\dagger(t)\} = \mathbf{A} \cdot \underbrace{E\{\mathbf{g}(t) \cdot \mathbf{g}^\dagger(t)\}}_{\mathbf{P}} \cdot \mathbf{A}^\dagger + E\{\mathbf{n_0}(t) \cdot \mathbf{n_0}^\dagger(t)\}, \tag{7.20}$$

where

$$\mathbf{P} = E\{\mathbf{g}(t) \cdot \mathbf{g}^\dagger(t)\} \tag{7.21}$$

represents the covariance matrix of the incoming signal. If all signals are uncorrelated, then $\mathbf{P}$ is a diagonal matrix:

$$\mathbf{P} = \begin{bmatrix} E\{|g_1(t)|^2\} & 0 & \ldots & 0 \\ 0 & E\{|g_2(t)|^2\} & \ldots & 0 \\ 0 & 0 & \ldots & E\{|g_Q(t)|^2\} \end{bmatrix}. \tag{7.22}$$

For uncorrelated noise at all antennas ($E\{\mathbf{n_0}(t) \cdot \mathbf{n_0}^\dagger(t)\} = \sigma^2 \mathbf{I}$), the covariance of the received signal (7.20) is thus given by

$$\mathbf{R} = \mathbf{A}\mathbf{P}\mathbf{A}^\dagger + \sigma^2 \mathbf{I} = \mathbf{R}_s + \sigma^2 \mathbf{I}. \tag{7.23}$$

The matrix

$$\mathbf{R}_s = \mathbf{A}\mathbf{P}\mathbf{A}^\dagger \tag{7.24}$$

is an $N_{\mathrm{ant}} \times N_{\mathrm{ant}}$ matrix with rank Q. This means that it has $N_{\mathrm{ant}} - Q$ eigenvectors associated with the eigenvalue of zero. For exactly such an eigenvector $\mathbf{u}_q$

$$\mathbf{A} \cdot \mathbf{P} \cdot \mathbf{A}^\dagger \mathbf{u}_q = \mathbf{R}_s \mathbf{u}_q = 0 \tag{7.25}$$

applies. Since $\mathbf{P}$ is positive definite, i.e. all its eigenvalues are greater than zero, it applies that

7.3 Angle Estimation Algorithms

$$\mathbf{A}^\dagger \mathbf{u}_q = 0. \tag{7.26}$$

This means that the $N_{\text{ant}} - Q$ eigenvectors $\mathbf{u}_q$ are orthogonal to all steering vectors $\{\mathbf{a}(\psi_1), \ldots, \mathbf{a}(\psi_Q)\}$ of the target directions. These eigenvectors form the matrix $\mathbf{U}_N$ with the dimension $N_{\text{ant}} \times (N_{\text{ant}} - Q)$. MUSIC seeks exactly those steering vectors that are orthogonal according to (7.26). Thus, the angular spectrum of MUSIC can be specified as

$$P_{\text{MUSIC}}(\psi) = \frac{\mathbf{a}^\dagger(\psi) \cdot \mathbf{a}(\psi)}{\mathbf{a}^\dagger(\psi) \cdot \mathbf{U}_N \cdot \mathbf{U}_N^\dagger \cdot \mathbf{a}(\psi)}. \tag{7.27}$$

It should be noted that this is not a true angular spectrum, but rather corresponds to a test for orthogonality. Therefore, (7.27) is referred to as a pseudo-spectrum. Whenever (7.26) is satisfied, the denominator in (7.27) becomes zero, causing a sharp peak to appear.

However, the covariance matrix $\mathbf{R}_s$ cannot be measured directly because it is always accompanied by the noise term in (7.24). Nevertheless, it can be estimated from $\mathbf{R}$. To do this, the number of targets in the channel is estimated using a model-order estimate. For this purpose, the covariance matrix $\mathbf{R}$ is decomposed. Since $\mathbf{R}_s \mathbf{u}_q = \lambda_q \mathbf{u}_q$ where λ_q corresponds to the eigenvalue belonging to $\mathbf{u}_q$, it follows that

$$\mathbf{R} \mathbf{u}_q = \mathbf{R}_s \mathbf{u}_q + \sigma^2 \mathbf{I} \mathbf{u}_q \tag{7.28}$$
$$= (\lambda_q + \sigma^2) \mathbf{u}_q. \tag{7.29}$$

This means that the eigenvectors of $\mathbf{R}_s$ are also eigenvectors of $\mathbf{R}$ with the eigenvalues $\lambda_q + \sigma^2$.

The covariance matrix $\mathbf{R}$, which can be estimated analogously to (6.18) from the last M measurements, is now decomposed into eigenvectors and eigenvalues:

$$\mathbf{R} = \mathbf{A} \cdot \mathbf{P} \cdot \mathbf{A}^\dagger + \sigma^2 \mathbf{I} = \mathbf{U} \cdot (\mathbf{\Lambda} + \sigma^2 \mathbf{I}) \cdot \mathbf{U}^\dagger \tag{7.30}$$

$$= \mathbf{U} \begin{bmatrix} \lambda_1 + \sigma^2 & 0 & \cdots & 0 & 0 & \cdots & 0 \\ 0 & \lambda_2 + \sigma^2 & \cdots & 0 & 0 & \cdots & 0 \\ \vdots & \vdots & \ddots & \vdots & \vdots & & \vdots \\ 0 & 0 & \cdots & \lambda_q + \sigma^2 & 0 & \cdots & 0 \\ 0 & 0 & \cdots & 0 & \sigma^2 & \cdots & 0 \\ \vdots & \vdots & & \vdots & \vdots & \ddots & \vdots \\ 0 & 0 & \cdots & 0 & 0 & \cdots & \sigma^2 \end{bmatrix} \mathbf{U}^\dagger. \tag{7.31}$$

The matrix $\mathbf{U}$ can therefore be decomposed into a signal part $\mathbf{U}_Q$ with Q columns, associated with the eigenvalues $\lambda_q + \sigma^2$, and a noise part $\mathbf{U}_N$ with $N_{\text{ant}} - Q$ columns, associated with the noise eigenvalues σ^2. All eigenvalues are either greater than or equal to σ^2 and can be attributed either to the signal or the noise component, whereby there are always Q eigenvalues greater than σ^2. From this assignment of the eigenvalues to either $\mathbf{U}_Q$ or $\mathbf{U}_N$, $\mathbf{U}_N$ is determined and then used in (7.27) to determine the angular spectrum.

In practice, however, this eigenvalue assignment or the estimation of the number of targets (model order) and thus the angular spectrum obtained from it, is prone to error. Especially in applications with many targets and a low SNR, the separation of the eigenvalues according to greater than or equal to σ^2 can be very difficult. If the eigenvalues are assigned incorrectly, then $\mathbf{U}_N$ is estimated incorrectly, which results in an incorrect angular spectrum. However, if the number of targets is known, $\mathbf{U}_N$ is determined directly and the MUSIC method proves to be much more robust. Therefore, high-resolution methods such as MUSIC are often combined with simpler beamformer approaches. In this case, the simpler beamformer approach serves to estimate the number of targets and to provide a rough angle estimate. With the knowledge of the number of targets, the high-resolution estimation method is applied and the rough estimate is used to check the plausibility of the high-resolution result. Furthermore, correlated noise at the receivers poses a problem to MUSIC since in this case $\mathbf{U}_N$ can no longer be determined correctly. Especially in the millimeter wave range, care must be taken to ensure that no correlated noise occurs, i.e., that the receivers operate uncoupled and independently of each other.

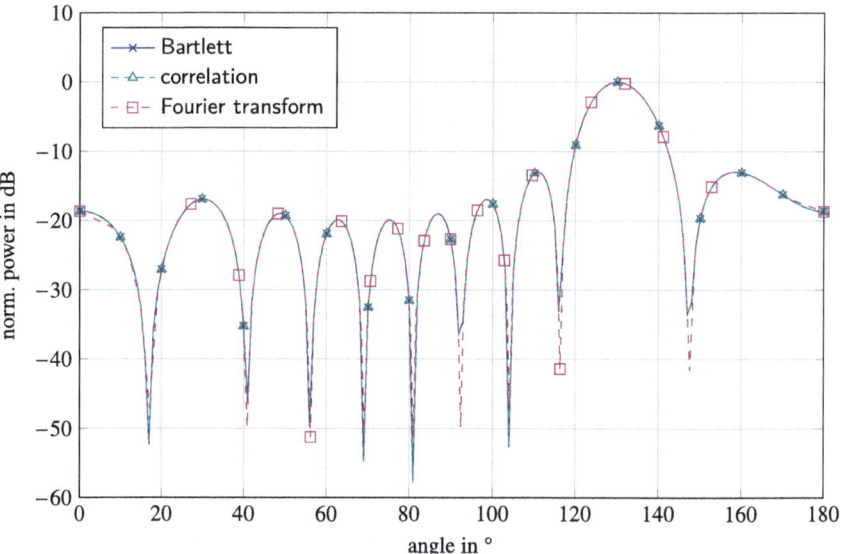

Fig. 7.1 Comparison of the correlation-based (*blue, solid*), Fourier-transform-based (*green, dashed*), and Bartlett beamforming (*magenta, dashed*) methods for the idealized case of no noise. The target is located at 130°. If there is no noise, all three methods converge

7.4 Comparison of the Methods

The angle estimation by correlation, by a Fourier transform and using Bartlett beamforming are to be compared first. The three approaches are identical if only one measurement is considered to determine the covariance matrix **R** and no noise occurs. This case is shown in Fig. 7.1. In practice, however, differences arise because measurements are limited by noise. Since the Bartlett beamformer minimizes the influence of noise by estimating the covariance matrix **R** based on many measurements, it is more robust to noise than the other two approaches. However, the influence of noise can also be reduced in the other two approaches by averaging. When using the correlation approach, it must be taken into account that each error or noise influence during the recording of the calibration matrix can lead to systematic errors.

Figure 7.2 shows the comparison of the Bartlett beamformer, the Fourier transform, the Capon beamformer and the MUSIC method. 1000 measurements were used for the evaluation of the covariance matrix, which explains the differences between the Fourier transform approach and the Bartlett beamformer. It can be observed that the Capon and MUSIC methods achieve significantly better resolutions and that their side lobe level is lower.

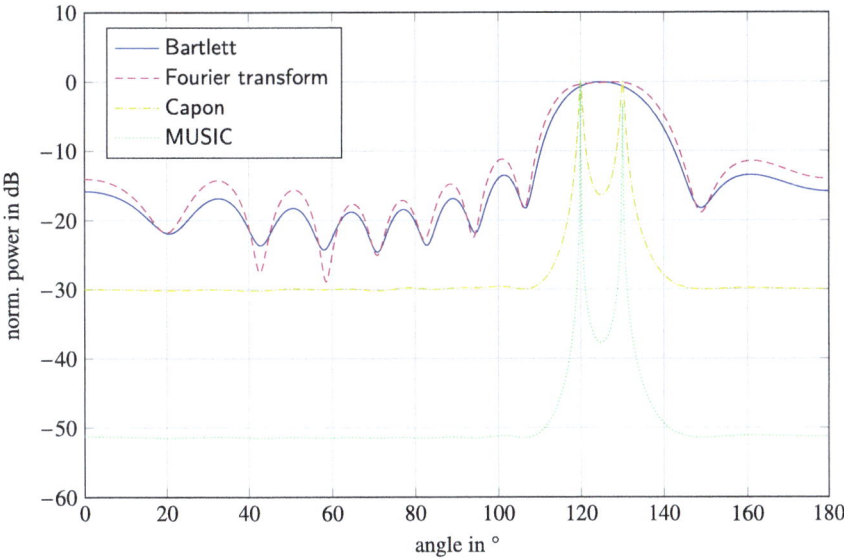

Fig. 7.2 Comparison of the Bartlett beamformer (*blue, solid*), Fourier-transform (*magenta, dashed*), Capon beamformer (*yellow, dash-dotted*), and MUSIC (*green, dotted*) angle estimation methods for an ULA consisting of 10 isotropic radiators with an antenna spacing of $\lambda/2$. Simulated angular spectra for two targets at 120° and 130°. MUSIC clearly shows the best performance, but it involves significantly more computing effort and is less robust

Open Access This chapter is licensed under the terms of the Creative Commons Attribution 4.0 International License (http://creativecommons.org/licenses/by/4.0/), which permits use, sharing, adaptation, distribution and reproduction in any medium or format, as long as you give appropriate credit to the original author(s) and the source, provide a link to the Creative Commons license and indicate if changes were made.

The images or other third party material in this chapter are included in the chapter's Creative Commons license, unless indicated otherwise in a credit line to the material. If material is not included in the chapter's Creative Commons license and your intended use is not permitted by statutory regulation or exceeds the permitted use, you will need to obtain permission directly from the copyright holder.

Chapter 8
MIMO Radars and Antenna Array Design

MIMO radars have multiple transmit and receive channels. Such multichannel radars have been used in different configurations for many decades. On the one hand, the detection probability can be increased compared to a single channel if the antenna spacing is very large and many receivers are used. The fusion of many channels leads to a lower probability of failure than a single channel if the channels have different fading characteristics. On the other hand, especially with small antenna spacings, in the range of $\lambda/2$, many channels can be used for classical transmit or receive beamforming. This approach leads to so-called *phased arrays*, which are described in Sect. 8.3.

However, the MIMO systems presented in this chapter address a completely different mode of operation where the numerous channels are used to make the receive aperture appear larger than it actually is. In this context this is called a *virtual aperture*. Since the virtual aperture is larger than the real aperture, this approach leads to radars with an improved angular resolution.

The main advantage of such MIMO radars is that only a small transmit and receive aperture has to be realized in hardware, but their common virtual aperture, which determines the angular resolution, is significantly larger. This advantage is so outstanding that almost all radars for angle estimation in the millimeter wave range are today realized as MIMO radars. In a MIMO design, the virtual aperture is at most twice as large as the real one, i.e., the sensor provides about twice the angular resolution of a radar without a MIMO concept at the same real or physical aperture size. Another advantage of the MIMO concept is that small antenna spacings, e.g. $\lambda/2$, can be realized in the virtual aperture, even if the individual radiators would not allow such spacings in the real aperture due to their size.

An important requirement for MIMO radars is that all transmit channels and all receive channels must be coherent, i.e. phase-locked to each other. This is the only way to create the virtual aperture, since the phase relationships between the antennas are evaluated. In addition, all transmit signals of each transmit channel must be orthogonal to each other to guarantee that the transmitted signals can be separated at the receiver.

This chapter begins with an introduction of the virtual aperture and the creation of orthogonal signals in Sects. 8.1 and 8.2, respectively. This is followed by a performance analysis of MIMO radars in Sect. 8.3, along with a description of how to design and evaluate MIMO radars in Sect. 8.4. Finally, Sect. 8.5 addresses the compensation of near field effects of large antenna arrays in the millimeter wave range.

8.1 Virtual Aperture

The virtual aperture can be introduced using different approaches, which are presented in the following. According to Chap. 6, the radiation pattern or field distribution resp. of an antenna array results from the Fourier transform of the current distribution on the antenna array, i.e. the array samples the current distribution. The signal path between the transmitter and the receiver results from the multiplication of the transmit and receive radiation pattern, as long as the LOS-component dominates the channel. Consequently, the signal path also results from the convolution of the current distributions of the transmit and receive arrays.

The illustration in Fig. 8.1 shows an single input multiple output (SIMO) antenna array and an incoming plane wave that causes a phase shift ϕ_m (6.12) between

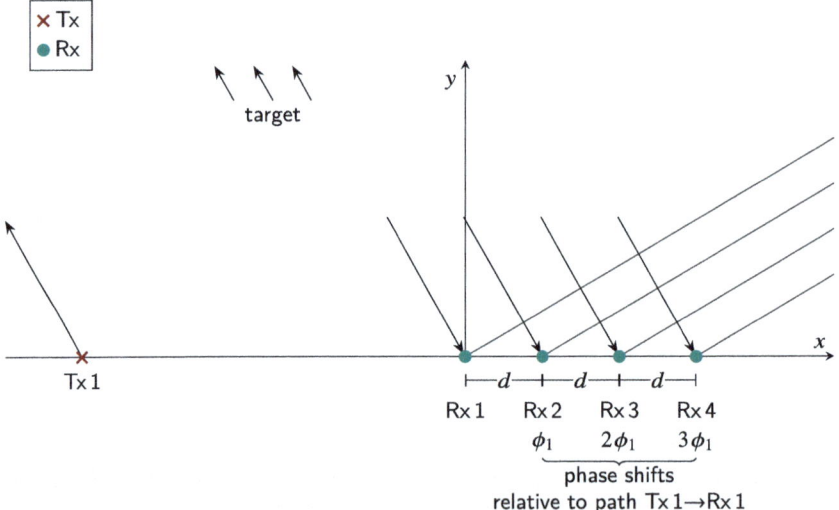

Fig. 8.1 Illustration of the phase shifts of the received signals at the receive antennas in a SIMO ULA system. The phases shifts are determined relative to the first receive antenna (*Rx 1*) and depend on the antenna spacings d

8.1 Virtual Aperture

the signals at the individual antennas, which are arranged at a physical spacing $d = x_{i-1} - x_i$. The corresponding steering vector (6.13)

$$\mathbf{a}(\psi) = \left[A_1(\psi) e^{j \frac{2\pi}{\lambda_0} d \cos(\psi)}, \ldots, A_{N_{\text{ant}}}(\psi) e^{j \frac{2\pi}{\lambda_0} N_{\text{ant}} d \cos(\psi)} \right] \quad (8.1)$$

follows.

If a second transmit antenna is set up at a spacing md from the first transmit antenna, a phase difference of $m\phi_1$ results. If a signal emitted by the second transmit antenna is received, there is a phase difference of $\phi_k + m\phi_1$, where ϕ_k describes the channel without antennas. Figure 8.2 shows this for $m = 4$.

Consequently, phase shifts occur in a MIMO system that are larger than those resulting from a pure consideration of the receive array. Here, the phase shifts between the transmit signals are added to the phase shifts between the signals at the receive antennas. The maximum phase shift occurring between the received signals results from the phase difference between the two outermost antennas in the array, whereby the phase difference of the transmitted signals is additive in a MIMO system. Therefore, the maximum phase difference in a MIMO system is greater than the maximum phase difference when considering just the receive array. This larger maximum phase difference corresponds to a larger aperture than the pure receive aperture. This apparently increased aperture of a MIMO system is referred to as the virtual aperture.

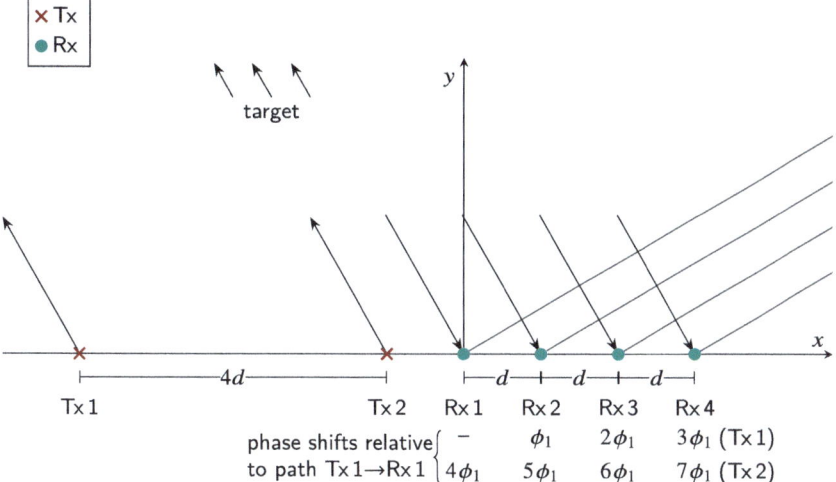

Fig. 8.2 Illustration of the phase shifts of the received signals at the receive antennas in a MIMO ULA system in which the spacing between the transmit antennas is exemplarily four times the distance between the receive antennas. The phase shifts are determined relative to the path from the first transmit antenna (*Tx 1*) to the first receive antenna (*Rx 1*).

Another possibility to introduce the virtual aperture is via the transmit and receive steering vectors. According to Fig. 8.2, the phase of the MIMO steering vector results from the sum of the phases of the transmit and receive steering vectors, i.e.

$$\mathbf{a}_{\text{mimo}} = \mathbf{a}_{\text{rx}}(\psi) \otimes \mathbf{a}_{\text{tx}}(\psi) \tag{8.2}$$

where $\otimes$ is the Kronecker product. Consequently, a virtual aperture with a maximum of $N_{\text{tx}} \cdot N_{\text{rx}}$ antennas can be constructed from N_{tx} transmit antennas and N_{rx} receive antennas. If the radiation patterns in the steering vectors are taken into account, the radiation patterns for the virtual antenna positions on the virtual aperture result from the product of the transmit and receive radiation patterns. It follows that the convolution of the transmit and receive aperture yields the virtual aperture, i.e. the spatial distribution function of the receive antennas convolved with the spatial distribution function of the transmit antennas yields the spatial distribution of the virtual aperture. Figure 8.3a and b shows two examples.

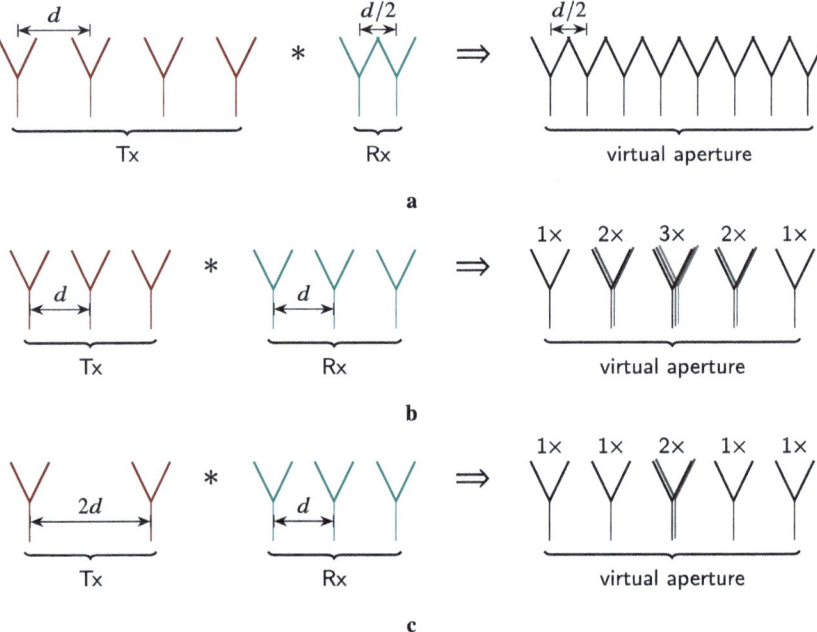

Fig. 8.3 Examples for the design and convolution of virtual ULAs with different transmit and receive configurations. The convolution (∗) of the spatial distribution function of the receiving antennas (*Rx*) with that of the transmitting antennas (*Tx*) results in the spatial distribution on the virtual aperture. **a** 4 × 2 array with equidistant transmit antenna spacing of d and receive antenna spacing of $d/2$. This configuration yields a dense virtual ULA with eight virtual antennas of spacing $d/2$ and total virtual aperture of $7d/2$. **b** 3 × 3 array with equidistant transmit and receive antenna spacings of d. This configuration yields a dense virtual ULA with five virtual antennas of spacing d and total virtual aperture of $5d/2$ where the second and forth virtual antenna positions are occupied twice while the middle antenna position is occupied even three times. **c** Optimized 2 × 3 array with transmitter spacing of $2d$ and receiver spacing of d. This configurations also yields five virtual antennas of spacing d and total virtual aperture of $5d/2$. However, compared to **b**, only the middle antenna is occupied twice

8.1 Virtual Aperture

Within the virtual aperture, antenna positions can be occupied multiple times, as shown in the example in Fig. 8.3b. Since such antenna positions are redundant, the design attempts to avoid them. Bistatic antenna configurations, in which the transmit and receive arrays differ, usually have significantly less redundancy than monostatic MIMO radars. Redundancy can also be reduced by reducing the number of antenna positions in uniform arrays. In the example in Fig. 8.3c, the transmit antenna in the center was removed compared to the example in Fig. 8.3b. This results in the same virtual aperture as in Fig. 8.3b but with less redundancy.

Basically, all of the above considerations also apply to two-dimensional arrays. For two-dimensional arrays, the antenna distribution function of the virtual aperture is obtained by convolving the two-dimensional distribution function of the transmit aperture with the two-dimensional function of the receive aperture, as shown in the example in Fig. 8.4.

The third possibility to introduce the virtual aperture is by describing the channel with transfer functions, which is similar to a description in terms of communications engineering. In this case, the channel between the transmit and receive antennas is described by transfer functions $H^{m_{rx}, m_{tx}}$ that are arranged in a $N_{tx} \times N_{rx}$ matrix $\mathbf{H}^{N_{rx}, N_{tx}}(\omega)$. The transfer function $H^{m_{rx}, m_{tx}}$ describes the sum of all propagation paths between the respective m_{tx}-th transmit and m_{rx}-th receive antennas. Each path is described by an amplitude and phase. Assuming that the LOS component dominates all other paths, the transfer matrix can be normalized and easily represented. This will be explained using the example in Fig. 8.5 with a transmit array of three antennas with spacing d and an identical receive array. The transfer function $H^{1,1}$ (*dark red*) between the first transmit and the first receive antenna is normalized to one. The transfer functions $H^{2,2}$ and $H^{3,3}$ (*dark red*) are identical, since the same LOS occurs and neither a path difference nor a phase shift to $H^{1,1}$ occurs at the transmitter or receiver. $H^{2,1}$ and $H^{3,2}$ (*turquoise*) are identical as well, but they are affected by a phase shift $-\phi$ compared to $H^{i,i}$, which occurs at the receiver. The transfer functions $H^{1,2}$ and $H^{2,3}$ (*yellow*) are also identical, but they have the same phase shift as $H^{2,1}$ and $H^{3,2}$ compared to $H^{i,i}$, only with reversed sign. $H^{1,3}$ (*magenta*) and $H^{3,1}$ (*blue*) experience twice the phase shift, but with opposite signs. The amplitudes

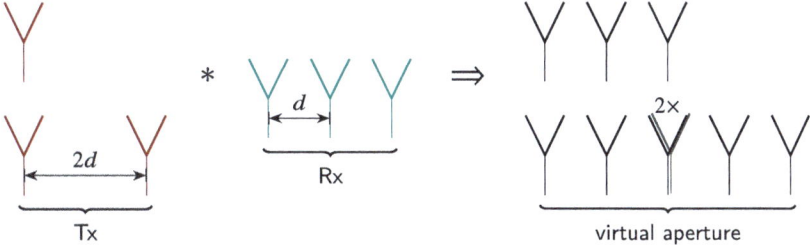

Fig. 8.4 Example for the convolution of virtual apertures for a two-dimensional MIMO aperture using a two-dimensional transmit array and a one-dimensional receive array. The spatial distribution function of the receive array is convolved with both rows of the transmit array which leads to a two-dimensional virtual array

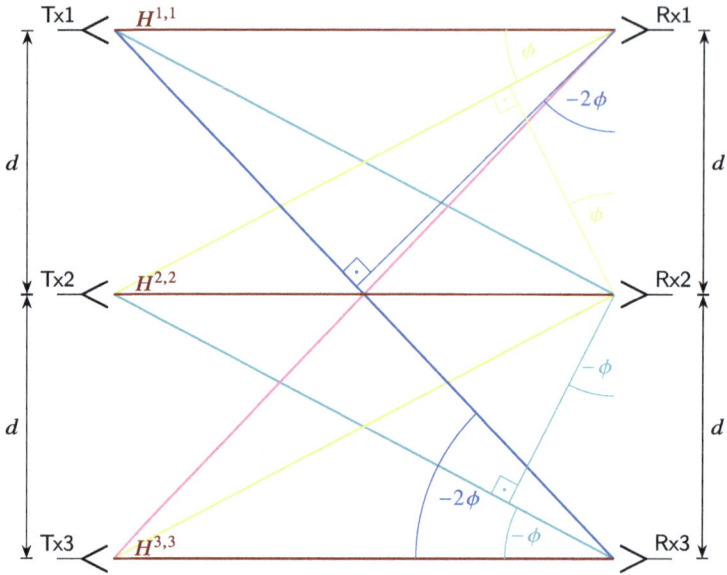

Fig. 8.5 Transfer functions in a 3×3 MIMO system. To cover all antenna combinations, nine transfer functions are necessary, of which only five are different. They differ by a phase shift due to the array configuration

of all transfer functions are approximately set to one. This results in the normalized channel matrix

$$\mathbf{H}^{3,3} = \begin{bmatrix} e^{j0} & e^{j\phi} & e^{j2\phi} \\ e^{-j\phi} & e^{j0} & e^{j\phi} \\ e^{-j2\phi} & e^{-j\phi} & e^{j0} \end{bmatrix} \quad \text{where} \quad \phi = md\cos\psi, \tag{8.3}$$

in which five different transfer functions occur. A radar system with one transmit antenna and five receive antennas, as shown in Fig. 8.6, is described by the identical five transfer functions

$$\mathbf{H}^{5,1} = \begin{bmatrix} e^{j2\phi} \\ e^{j\phi} \\ e^{j0} \\ e^{-j\phi} \\ e^{-j2\phi} \end{bmatrix}. \tag{8.4}$$

This shows that a 1×5-MIMO system can be calculated based on the 3×3-MIMO system, which, according to Fig. 8.6, has an aperture twice as large as that of the 3×3-MIMO system. The resolution is approximately doubled as a result.

8.2 Generation of Orthogonal Signals

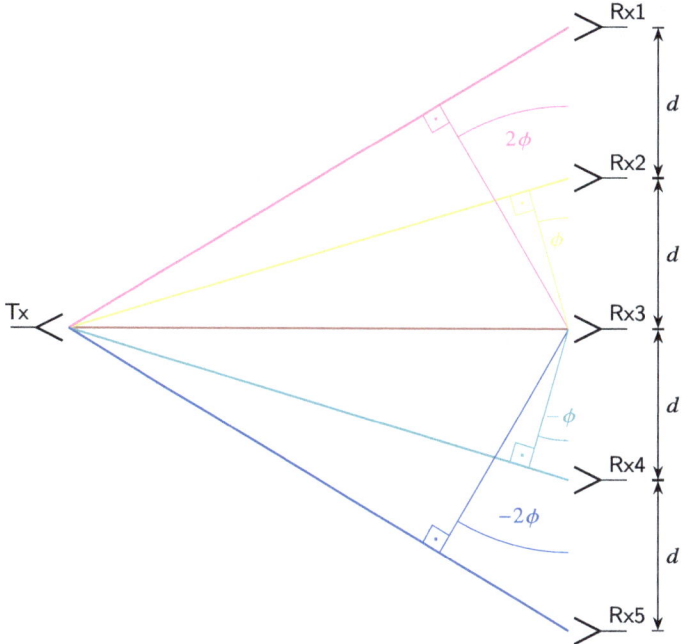

Fig. 8.6 Transfer functions in a 1×5 SIMO system. The same transfer functions occur here as in the 3×3 MIMO system, which shows the equivalence of the two systems

8.2 Generation of Orthogonal Signals

In the case of a MIMO radar, each transmit antenna emits a different signal. In order for the transmit signals to be distinguishable at the receiver, they must be orthogonal to each other. In order to achieve this, there are essentially three different approaches, which are discussed in this section.

8.2.1 Time Division Multiplexing

The simplest approach to generate orthogonal signals is time-division multiplexing (TDM). In this case, the antennas are switched through one after the other so that only one antenna is transmitting at a time. TDM is particularly suitable for signals that have a temporal structure, such as a chirp sequence signal. The individual chirps are distributed among the transmit antennas, as shown in Fig. 8.7, i.e. after each chirp, the system switches to the next antenna.

Although the method is used very frequently, especially with chirp sequence radars, it has some disadvantages. According to (5.37), the chirp repetition rate determines the unambiguous Doppler shift or the unambiguously measurable velocity. If

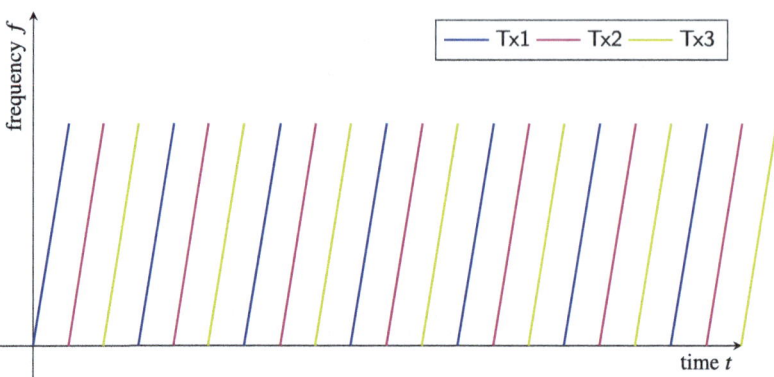

Fig. 8.7 The TDM-MIMO multiplexing scheme using the example of a chirp sequence signal with three transmitters. The ramps are emitted one after the other by the different transmitters and are orthogonal to each other due to the time offset

the repetition period of the chirps per transmit antenna is reduced with TDM-MIMO by a factor corresponding to the number of transmit antennas N_{tx}, the maximum unambiguous velocity reduces likewise to

$$v_{\text{uaTDM}} = \frac{v_{\text{ua}}}{N_{\text{tx}}}. \tag{8.5}$$

Thus, the maximum unambiguous velocity v_{uaTDM} is reduced by a factor of N_{tx} compared to the unambiguous velocity of a SIMO or single input single output (SISO) system.

Furthermore, when chirp sequence modulation and TDM are applied a relevant coupling of the velocity and angular information occurs, since the phase differences between the signals at the virtual antenna positions, from which the angle is determined, are recorded at different times. The angle-dependent phase difference $\phi = md\cos(\psi)$ is extended by a Doppler-dependent term. For the m_{tx}-th transmitter, the following applies

$$\phi_{m_{tx}} = m_{tx} d \cos(\psi) + 2\pi f_D T \frac{m_{tx} - 1}{N_{tx}}. \tag{8.6}$$

This means that any phase differences detected by all receivers using the same transmitted signal are not affected by the Doppler shift. However, after switching to the next transmit antenna, there is a phase change that depends on the Doppler shift. This results in a block-wise Doppler dependent phase shift of the antenna elements of the virtual aperture.

However, the velocity-angle coupling can easily be compensated for when calculating the Doppler spectrum by correcting the angle-dependent phase term in the velocity-DFT, as described in more detail in [1]. Instead of (5.34), the following

8.2 Generation of Orthogonal Signals

modified DFT is calculated:

$$\mathcal{V}[\eta] = \sum_{l=0}^{L-1} e^{j\phi_{m_{tx}}} e^{j2\pi f_D l T_{rep}} e^{-j2\pi \frac{\eta}{L}\left(l+\frac{m_{tx}-1}{N_{tx}}\right)}. \tag{8.7}$$

As a result, only the angle-dependent phase information remains in the R-v matrix, which is used for subsequent angle estimation of the targets.

8.2.2 Frequency Division Multiplexing

To create orthogonality between the different transmit signals, different frequencies are assigned to the individual transmit antennas or transmit signals for frequency-division multiplexing (FDM). This approach is particularly suitable for multicarrier methods such as OFDM. Here, the individual subcarriers are distributed to the different transmit antennas. Since the subcarriers are all likewise orthogonal, the individual transmit signals of the individual transmit antennas are orthogonal to each other. The disadvantage of this method is that the range unambiguity is reduced by a factor corresponding to the number of transmit antennas N_{tx}:

$$R_{\text{uaFDM}} = \frac{R_{\text{ua}}}{N_{tx}}. \tag{8.8}$$

This is similar to the reduction of the unambiguous velocity with TDM. Also with analog radars like chirp sequence FDM can be used, however, only at the expense of a higher bandwidth of the resulting beat signal. The ramps of the individual transmit signals are shifted relative to each other in the frequency domain to such an extent that only the mixing with one of the transmit ramps at a time generates a beat signal in the receiver being within the filter bandwidth of the anti-aliasing filter of a corresponding SISO system, as shown in Fig. 8.8. This means that the frequency spacing of the individual chirps must be at least twice the bandwidth of the SISO anti-aliasing filter. This results in a bandwidth of the chirp sequence FDM beat signal of

$$f_{\text{max,FDM}} = 2 N_{tx} f_{\text{max},1} \tag{8.9}$$

and the bandwidth of the RF signal is given by

$$\Delta f_{\text{FDM}} = \Delta f_1 + 2 N_{tx} f_{\text{max},1}. \tag{8.10}$$

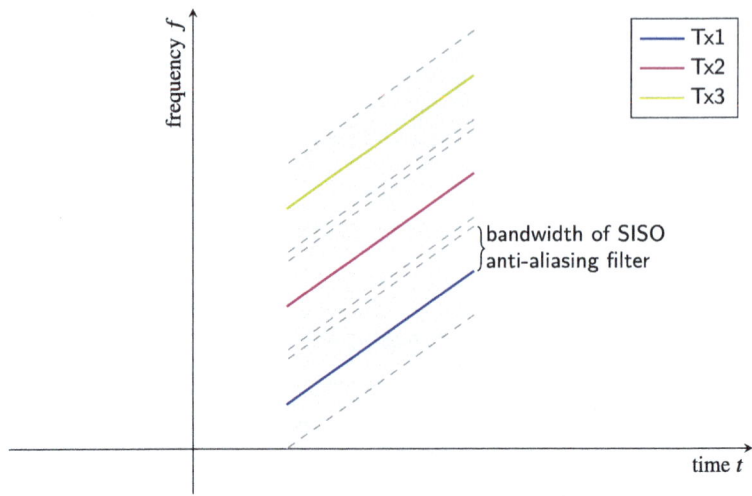

Fig. 8.8 The FDM-MIMO-procedure using the example of a chirp sequence signal with three transmitters

8.2.3 Code Division Multiplexing

The use of orthogonal codes to create orthogonality between transmit signals is the basis of code-division multiplexing (CDM). This means that each transmit signal is assigned and modulated with a unique code. This approach is particularly suitable for PMCW modulation, since this is based on the use of codes with good autocorrelation properties anyway. However, if several orthogonal transmit signals are to be generated, the cross-correlation properties of the codes are crucial. Thus, both correlation properties can be important in radar operation. However, most codes have significantly better autocorrelation than cross-correlation properties. A large number of codes are known from communications engineering that are suitable for this purpose. Of particular note are the maximum length, Gold and Kasami sequences. The autocorrelation and cross-correlation properties of which are illustrated in Figs. 4.17 and 4.18.

Unlike in communications technology, MIMO radars often do not use different, mutually orthogonal codes for the various transmitters, but rather use the same code for all transmitters, whereby the signals of the transmitters are time-shifted relative to each other. This time shift of the code for the individual transmit signals must be greater than the propagation time of the signal in the channel. This allows the autocorrelation properties to be used when correlating at the receiver, in contrast to the cross-correlation properties when different codes are used. This means that significantly larger target dynamics can be mapped, since in PMCW radar the maximum target dynamics are given by the side lobe spacing of the correlation function, as explained in Sect. 4.2.2. However, by using time-shifted copies of the same signal between the transmitters, the maximum unambiguous range is reduced, similarly to

FDM-MIMO. I.e. for N_{tx} transmitters, the maximum time shift between any two transmitters is T/N_{tx}; Thus, the maximum unambiguous range reduces by a factor corresponding to the number of transmit antennas:

$$R_{uaCDM} = \frac{R_{ua}}{N_{tx}}. \tag{8.11}$$

8.3 Phased Arrays and Performance Analysis for MIMO Radars

MIMO systems in the millimeter wave range are primarily used to increase the angular resolution of a radar sensor without increasing the real aperture. As a result, they have become established in a wide range of applications for imaging radar sensors.

In contrast, so-called phased arrays are traditionally used in radar applications in which beamforming is to be carried out on the transmit and receive sides. In a phased array, a single transmit signal is distributed in-phase to all transmit antennas so that the main beam direction of the transmitter can be steered in the direction of the targets. By focusing the signal using multiple transmit antennas, a narrow main lobe is obtained, which consequently illuminates only a small area, as shown in Fig. 8.9a. The targets in the illuminated area reflect the signal in the direction of the receiver, where beamforming, i.e. a phase-correct addition of the signals from all receive channels, is performed.

With MIMO techniques, on the other hand, an orthogonal signal is transmitted via each transmit antenna, i.e. there is no focused main lobe and a significantly larger area is illuminated, as can be seen in Fig. 8.9b. At the receiver, the signals can be separated due to the orthogonality and beamforming can be performed.

These explanations initially suggest that classic phased arrays lead to a higher SNR compared to MIMO systems. However, this must be examined more closely. Different integration gains (see also Sect. 5.1) are included in the SNR of a multi-channel radar:

$$\text{SNR} \propto I_{tx} I_{rx} I_f I_t I_S. \tag{8.12}$$

The gains I_f and I_t describe the integration gain in fast and slow time dimension, respectively. I_{tx} and I_{rx} are the transmit and receive integration gains respectively the number of transmit and receive channels per signal. I_S describes the number of different transmit signals. To compare the achievable gains for MIMO and phased array systems, initially the same system requirements are assumed, such as the same number of transmit and receive channels, as well as the same integration gain over fast and slow time; Thus, the factors I_{rx}, I_f and I_t are initially irrelevant. Furthermore, it is assumed that the same power is radiated in both cases, regardless of how many transmitters are active at the same time. Then, the product of $I_{tx} \cdot I_S$ is constant for

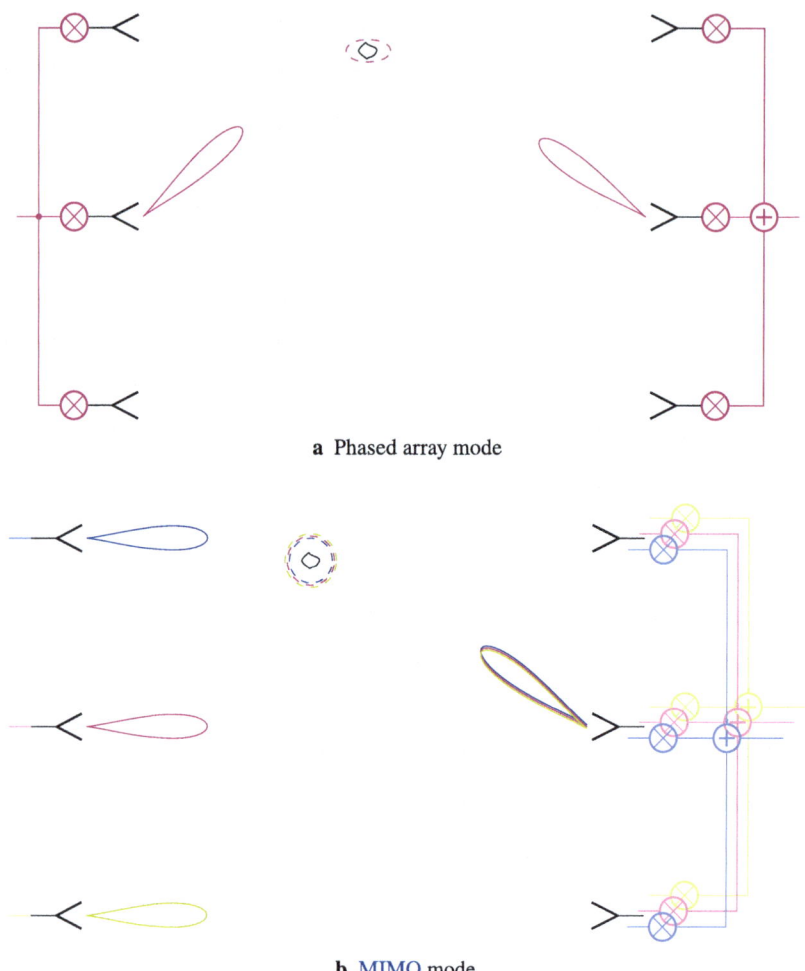

Fig. 8.9 Comparison of **a** a classic phased array with only one transmit signal with **b** a MIMO radar, in which the number of mutually orthogonal transmit signals corresponds to the number of transmit antennas

the two operating modes, phased array and MIMO. For FDM- and CDM-MIMO, it holds that $I_{tx} = 1$ and $I_S = N_{tx}$. For phased arrays, on the other hand, $I_{tx} = N_{tx}$, but $I_S = 1$. Thus, both modes of operation eventually lead to the same integration gain and consequently the same SNR. The prerequisite for this is that with MIMO all transmitted signals are integrated at the receiver, which is possible due to the orthogonality.

The operating modes only differ with regard to the SNR if irregular sampling is used, which is the case for TDM-MIMO with the same measurement duration, i.e. integration duration of the signals. In this case, only one transmit antenna emits a signal at a time, i.e. $I_{tx} = 1$ and $I_S = 1$, whereby a lower SNR is achieved for the same measurement duration for TDM-MIMO. However, if for TDM-MIMO the integration is N_{tx} times longer, so that the same transmit energy as with a phased array is emitted into the channel, both achieve the same SNR. In this case, I_t increases by the factor N_{tx}, which again achieves the same SNR as with a phased array.

Phased arrays and MIMO systems achieve the same SNR under the same signal conditions. With TDM-MIMO, the duration of the measurement must be increased by the number of transmitters due to the time required to switch the transmitters in order to get the same SNR.

8.4 Array Design and Evaluation

To assess the performance of an array, various evaluation criteria are used in the design of antenna arrays for millimeter-wave radars. The interface to the transmitter and receiver is classically described by the input impedance or the reflection factor, respectively. The description of the air interface is more diverse. Here, the FoV is used, as described in Sect. 6.1. In addition, the array ambiguity function plays an important role in evaluating the antenna pattern of an antenna array. The array ambiguity function is defined as

$$\chi(\psi_i, \psi_j) = \frac{\left|\mathbf{a}^\dagger(\psi_i)\mathbf{a}(\psi_j)\right|}{\|\mathbf{a}(\psi_i)\|\|\mathbf{a}(\psi_j)\|}. \tag{8.13}$$

It corresponds to the normalized correlation of the steering vectors of the two angles ψ_i and ψ_j. Due to the normalization, each element $\chi(\psi_i, \psi_j)$ (8.13) of the matrix χ takes values between zero and one and describes the similarity of the steering vectors $\mathbf{a}(\psi_i)$ and $\mathbf{a}(\psi_j)$ of the two angles ψ_i and ψ_j. If the steering vectors of the two angles are identical, the correlation is one, i.e. $\chi(\psi_i, \psi_j) = 1$. Figure 8.10 shows examples of the array ambiguity function for various antenna arrays.

The matrix χ with elements (8.13) can be interpreted as follows: One of the two angles ψ_i, ψ_j is interpreted as the angle at which the waves or radar signals incident on the array. The other angle is the angle that the antenna array interprets as the angle of arrival. Thus, the antidiagonal of the matrix (diagonal from the top right to the bottom left) must always be set to one so that real angles of arrival correspond to the measured angles. However, if high values close to one occur beyond the antidiagonal, there is a high correlation between the real angle of arrival and another, wrong angle, under which the radar target only appears to be. A high correlation off the antidiagonal indicates a high probability of confusing two angles of arrival or creating a so-called ghost target. High values off the antidiagonal result in high side lobes. Moreover, in case of ones off the diagonal, ambiguities or grating lobes emerge. A well-designed antenna array has an ambiguity function that has a narrow antidiagonal with values

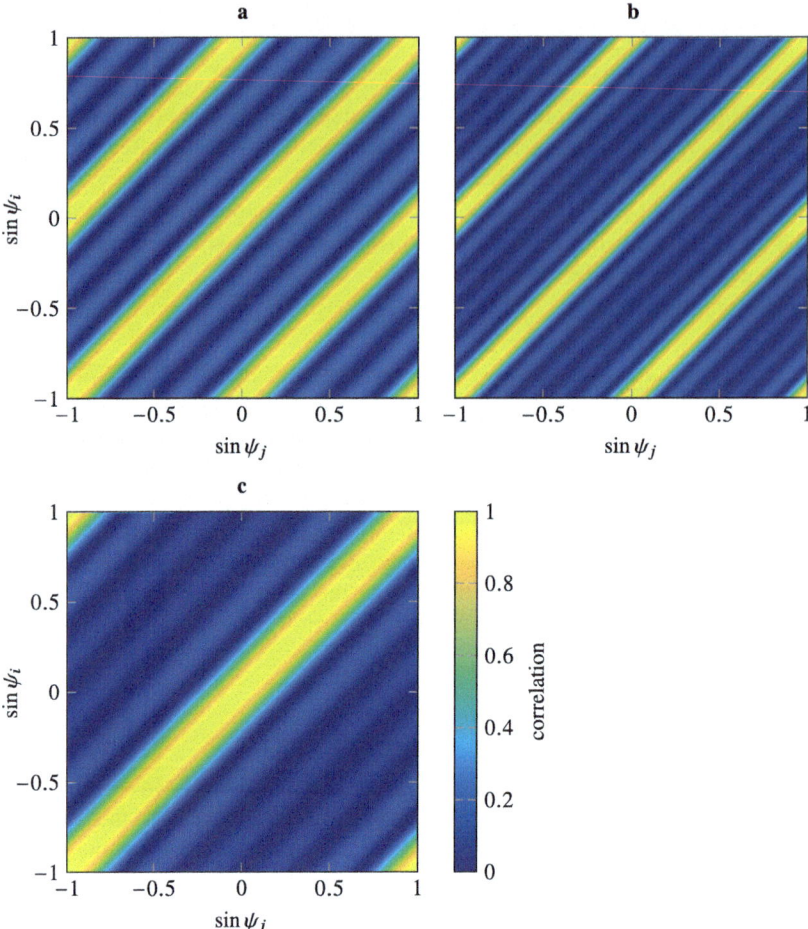

Fig. 8.10 Ambiguity function for ULAs with **a** four elements and $d = \lambda$, **b** six elements and $d = \lambda$, and **c** six elements and $d = \lambda/2$. The narrower the antidiagonal, the better the resolution of the array. Peaks beyond the antidiagonal indicate a high probability of angular errors or ambiguities

of one and otherwise takes the lowest possible values. If the antidiagonal is narrow, a high angular resolution is achieved.

In the ambiguity function χ (8.13), the array geometry, i.e., the spatial arrangement of the antennas, is primarily mapped via the steering vector. However, the complex-valued radiation pattern can be taken into account in the steering vector, especially if different antennas are used in the array. It should be noted that this type of ambiguity function strictly speaking only applies to the single-target case, because the steering vectors of several targets can superpose arbitrarily. However, this form of the array ambiguity function has proven itself in a large number of practical cases, especially when targets are primarily separated by range and velocity, so that only one target per R-$v$cell needs to be considered for angle estimation. The ambiguity function for two-dimensional angular radars is a four-dimensional function, since angle pairs have to be mapped onto each other, see [2].

Using the array ambiguity function and taking the FoV into account, an antenna array is designed from various aspects, as follows.

To achieve a high angular resolution, the aperture must be as large as possible according to (3.18) and (6.24). However, this requires a large number of receive antennas or channels if the aperture is to be fully sampled, i.e. in the $\lambda/2$ grid. To avoid grating lobes or ambiguities, the antenna spacing $\lambda/2$ must be maintained at least once in the aperture, as shown in (6.33). If this condition is violated, ambiguities in the array ambiguity function cannot be avoided. If the aperture is sampled in a uniform $\lambda/2$ grid, this is referred to as a fully populated array. For reasons of cost and space, it is often attempted to sample the largest possible aperture with only a few antennas, i.e. the $\lambda/2$ grid is not maintained. If at least once the $\lambda/2$ spacing is maintained, no ambiguities are to be expected, but the side lobes increase, i.e. the array ambiguity function shows many high values off the antidiagonal. By intelligent placement of the antennas on the aperture, these side lobes can be kept as low as possible in the array ambiguity function. For this purpose, optimization methods are used in modern array design, where the optimization goal is to achieve low values outside the antidiagonals [2]. Most often, so-called genetic optimization methods are used here.

Heavily sparse arrays with antenna spacings greater than $\lambda/2$ are often sensitive to position tolerances. This means that even small position errors can lead to a significant deterioration in the array ambiguity function. This is especially critical in the millimeter wave range, since etching and manufacturing tolerances lead to significant position tolerances in relation to the wavelength. In this case, a tolerance analysis must be taken into account in the optimization of the antenna array.

A further disadvantage of sparse, non-uniform arrays is the limited possibility of using window functions in angle estimation to suppress the side lobes. The familiar window functions, as introduced in Sect. 5.2, can only be used effectively for uniform antenna arrays (ULA or URA). For sparse arrays, the signal must be resampled or a uniform sample reconstruction must be performed in an additional processing step before a window function can be applied.

8.5 Compensation of Near-Field Effects

In many applications in the area of millimeter waves, large antenna arrays are used in relation to the wavelength for relatively small target ranges. This combination of large arrays and small ranges often results in the far-field condition according to (1.1), which is the basis for all array considerations, no longer being met. When using MIMO systems, this is further exacerbated because the far-field distance R_{FF} not only results from the size of the virtual aperture, but also from the distance between the transmit and receive apertures, among other parameters. This is due to

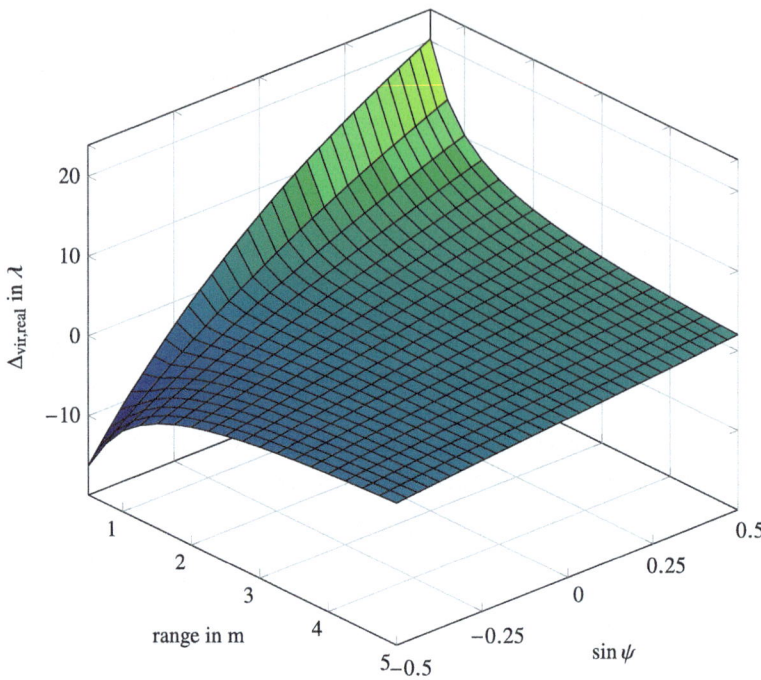

Fig. 8.11 Difference $\Delta_{\text{vir,real}}$ of the virtual near-field antenna position from the virtual far-field antenna position for a monostatic aperture of size 50 wavelengths at 160 GHz, [3]

the assumption of a plane wave made when deriving the virtual aperture. The plane wave extends over all virtual transmit and receive combinations and all positions of the transmit and receive antennas, respectively. In violation of the far-field condition, a systematic error occurs both during calibration of an antenna array and during angle estimation in normal radar operation. This error can lead to a reduction in the SNR, a deterioration in angular resolution or to a false number of targets. In applications with low requirements, this error can be ignored if necessary. However, as requirements increase, the error must be corrected to achieve the optimum measurement quality achievable with an array.

Figure 8.11 shows the differences in the virtual antenna positions between a spherical wave and a plane wave when incident on an aperture. The difference in path length between a spherical wave and a plane wave results in a phase error that increases as the target distance decreases. Assuming plane waves and respecting the far-field condition, the phase at a virtual antenna position is obtained from the signal phases of the transmit and receive paths

$$\Delta\phi_{\text{vx,FF}} = \Delta\phi_{\text{tx,FF}} + \Delta\phi_{\text{rx,FF}}. \tag{8.14}$$

8.5 Compensation of Near-Field Effects

If, on the other hand, the far-field condition is violated, i.e. the assumption of a plane wave is no longer justified, then $\Delta\phi_{\text{vx,FF}}$ should be corrected. The correction term depends on the range and azimuth of the target. A target at range R from the antenna array at angle ψ can be estimated by

$$\mathbf{R} = \begin{pmatrix} -R\sin(\psi) \\ R\cos(\psi) \end{pmatrix}. \tag{8.15}$$

This can be used to define a short-range steering vector based on the assumption of the incidence of a spherical wave as

$$\mathbf{a}_{\text{rx,NF}} = \left(e^{j\frac{2\pi}{\lambda}\left(|\mathbf{R}-\mathbf{x}_{\text{rx},1}|-|\mathbf{R}|\right)}, \ldots, e^{j\frac{2\pi}{\lambda}\left(|\mathbf{R}-\mathbf{x}_{\text{rx},N}|-|\mathbf{R}|\right)} \right) \tag{8.16}$$

$$= \left(e^{j\Delta\phi_{\text{rx,NF},1}}, \ldots, e^{j\Delta\phi_{\text{rx,NF},N}} \right). \tag{8.17}$$

This steering vector applies to the receive case as defined. For MIMO systems, it is defined analogously for the transmit case, since the total phase progression according to (8.14) results from the transmit and the receive case.

To correct a steering vector measured in the near field for the incidence of a plane wave, the Hadamard product ($\odot$) applies

$$\mathbf{a}_{\text{rx,FF}} = \mathbf{a}_{\text{rx,NF}} \odot \mathbf{a}_{\text{rx,korr}}. \tag{8.18}$$

The resulting correction vector is

$$\mathbf{a}_{\text{rx,korr}} = \left(\mathbf{a}_{\text{rx,NF}} \odot \left(\mathbf{a}_{\text{rx,FF}}^* \right) \right) \tag{8.19}$$

$$= \left(e^{j\Delta\phi_{\text{rx,NF},1}-\phi_{\text{rx,FF},1}}, \ldots, e^{j\Delta\phi_{\text{rx,NF},N}-\phi_{\text{rx,FF},N}} \right). \tag{8.20}$$

This correction vector can be used not only to convert a steering vector measured in the near field to the far field, but also to convert calibration parameters recorded in the near field to the far field. Only calibration steering vectors for the far field are independent of the range and angle of arrival of the calibration target. For this reason, it is essential for most applications to convert calibration data recorded in the near field to the generally valid case of the far field, provided that the targets are also in the far field. For targets in the near field, the calibration parameters must always be converted to the correct near-field range.

References

1. J. Bechter, F. Roos, C. Waldschmidt, Compensation of motion-induced phase errors in TDM MIMO radars. IEEE Microw. Wirel. Compon. Lett. **27**(12), 1164–1166 (2017). ISSN: 1558-1764. https://doi.org/10.1109/lmwc.2017.2751301
2. A. Di Serio, P. Hugler, F. Roos, C. Waldschmidt, 2-D MIMO radar: a method for array performance assessment and design of a planar antenna array. IEEE Trans. Antennas Propag. **68**(6), 4604–4616 (2020). ISSN: 1558-2221. https://doi.org/10.1109/tap.2020.2972643
3. A. Dürr, Koppelbare Radarsensoren zur hochauflösenden Bildgebung bei 150 GHz. German, Ph.D. Dissertation, Universität Ulm, 2022. https://doi.org/10.18725/OPARU-46249

Open Access This chapter is licensed under the terms of the Creative Commons Attribution 4.0 International License (http://creativecommons.org/licenses/by/4.0/), which permits use, sharing, adaptation, distribution and reproduction in any medium or format, as long as you give appropriate credit to the original author(s) and the source, provide a link to the Creative Commons license and indicate if changes were made.

The images or other third party material in this chapter are included in the chapter's Creative Commons license, unless indicated otherwise in a credit line to the material. If material is not included in the chapter's Creative Commons license and your intended use is not permitted by statutory regulation or exceeds the permitted use, you will need to obtain permission directly from the copyright holder.

Part III
Radar Hardware

Chapter 9
Hardware and Technology

Radar sensors in the millimeter wave range usually consist of an antenna system, a radar circuit, a digital processor for signal evaluation and a power supply. The radar circuit is monolithically integrated and is referred to as MMIC. The antenna system used is either an array consisting of many individual antennas, as described in Chap. 6, or a single antenna, which is often designed as a primary radiator together with beamforming lenses. This chapter focuses on the two components MMIC and antenna, as the processing unit and power supply are usually available as standard components.

9.1 System Partitioning

In the millimeter wave range, special boundary conditions must be taken into account due to the high frequency, which leads to specific system partitioning concepts. At frequencies below approx. 100 GHz, the antenna system and the MMIC are often set up geometrically separated from each other. On the one hand, this is because transmission line losses between the MMIC and the antenna system are still acceptable and the tolerance requirements for transmission lines or line transitions, such as wire bond transitions, can still be realized in various technologies. On the other hand, antennas that are integrated directly into the MMIC are large, so that the required silicon area would lead to high costs.

For radar sensors above approx. 100 GHz, however, it is hardly possible to separate the antennas and the MMIC spatially due to the losses and high tolerance requirements for the high-frequency structures, such as transitions or other passive components. In this case, the antennas are integrated directly into the MMIC, or alternatively integrated together with the MMIC in a housing or package. In this case, both are placed so close to each other that the signal is fed into the antenna electromagnetically or, as a special solution, via very short transmission lines, i.e. by means of galvanic coupling from the MMIC. Due to these boundary conditions, various partitioning concepts have become established, which, however, are not always differentiated from each other in their specific designs, but rather complement each other and are combined.

MMIC and antenna system detached If large antenna arrays are to be used for beamforming or angle estimation, they are usually set up detached from the MMIC on a carrier. This approach is only common up to frequencies of approx. 100 GHz. The antenna array is then constructed as an independent component, e.g. using waveguide technology or printed circuit board (PCB) technology. Implementation variants for this follow in Sect. 9.3.

Antenna in package (AiP) If the transmission line losses between the antenna and MMIC are to be kept low, the antenna can be integrated directly into the package of the MMICs. However, since the package of an MMIC is usually small compared to an array for angle estimation, normally only one or very few antennas are integrated into the package as primary radiator for a lens. Since the antenna interacts with the entire package, the package with its RF characteristics and tolerances must be taken into account in the design. Package materials with high losses are not suitable for AiP concepts. Packages with integrable antennas are presented in Sect. 9.2.3.

Antenna On Chip (AoC) The AoC concept refers to the integration of the primary radiator into the MMIC. Due to the size of the antennas, this only makes sense above approx. 100 GHz. However, the high losses in the silicon bulk are problematic with these AoC approaches, so special techniques should be used to avoid the losses. Implementation variants for this concept are discussed in Sect. 9.3.3.

System on Chip (SoC) The increasing use of CMOS technologies for the MMIC makes it possible to integrate the processing unit and other logic units together with the millimeter wave front end in an integrated curcuit (IC). This concept is implemented particularly in the automotive sector due to the high production volumes of radar sensors. Such system partitions are referred to as SoCs and achieve the highest level of integration available for radar sensors.

9.2 Assembly and Interconnect Technology for Antennas and MMICs

Assembly and interconnect technology (AIT) in the millimeter wave range has very different levels of technical maturity depending on the frequency range. While various standard solutions have become established as packages in the frequency range below approx. 100 GHz and special millimeter wave PCBs are used, this is not the case in the frequency range above 100 GHz. At these high frequencies, very different approaches and solutions can be found in the literature, none of which have yet become established for high-volume applications. PCBs are also hardly usable in this frequency range.

9.2 Assembly and Interconnect Technology for Antennas and MMICs

Table 9.1 Selection of some materials with permittivity ϵ_r and loss angle $\tan \delta$ frequently used in the millimeter wave range of 140–220 GHz according to [[1], Table A.1]

Material		ϵ_r	$\tan \delta$
Substrate material	RO3003	2.99	0.0015
	RO5880	2.36	0.0015
	RT5870	2.51	0.0034
	Taconic 605	2.30	0.0024
Plastic material	PTFE	2.06	0.0011
	HDPE	2.24	0.0012
Mold compound	HIK powder	10.15	0.0039
	L900HF	9.11	0.0024
	PP300	2.95	0.0010
	PP500	5.00	0.0012
	PP600	5.95	0.0017

Section 9.2.1 begins with some general considerations on the selection of materials and technologies. The topics millimeter wave PCBs, packages and transitions follow in Sects. 9.2.2, 9.2.3, and 9.2.4.

9.2.1 Material and Geometry Requirements

For the design of structures such as antennas, lenses, PCBs, waveguides, etc., as well as for structures that are penetrated by fields such as packages or parts thereof, only materials with low attenuation and a permittivity ϵ_r that is stable over the frequency range can be used in the millimeter wave range. For dielectric materials with a low conductivity ρ, the loss angle results from

$$\tan \delta = \frac{1}{\omega \rho \epsilon_r \epsilon_0}. \tag{9.1}$$

The material parameters of many materials are often not known in the millimeter wave range and the measurement of the parameters is very complex and error-prone. For this reason, some materials are listed as examples in Table 9.1.

Copper or gold coatings are often used for the design of current-carrying metallic structures, depending on the required conductivity. In the case of metallic structures, the surface roughness is of decisive importance for the losses due to the low skin penetration depth in the millimeter wave range. Roughness is understood as the root mean square (RMS) value of the deviation from a smooth surface. A very high roughness can double the surface resistance. With a roughness of one skin depth (approx. 0.2 µm at 100 GHz in copper), the surface resistance already increases by 60% compared to an ideally smooth surface, with a roughness of half a skin depth by

20%. Furthermore, the manufacturing tolerances are of great importance for metallic structures, as explained in the following section for components in PCB technology as an example.

9.2.2 Millimeter Wave Printed Circuit Boards

In the frequency range up to approx. 100 GHz, circuits and antennas are often built on printed circuit boards (PCBs). These are easily available and many millimeter wave components are developed with corresponding transmission line types and transitions for use on PCBs. Despite the high maturity of the technology, PCBs up to approx. 100 GHz are very different from PCBs for lower frequencies.

A key distinguishing feature is the need for special millimeter wave substrates that have a defined behavior in the millimeter wave range, i.e. a specified permittivity ϵ_r and a low loss angle tan δ. Ordinary FR4 substrates are ruled out in the millimeter wave range because they have high losses and a strongly fluctuating permittivity. Some example values for the permittivity and loss angle of typical substrate materials are listed in Table 9.1. High-frequency PCBs usually consist of several layers of FR4 for the low-frequency electronics or signal routing as well as one or more layers of millimeter wave substrate on one side for the millimeter wave circuit or antenna. Millimeter wave substrates often have material properties that must be taken into account when designing and manufacturing PCBs. The substrates are often mechanically soft, which is why a mechanically symmetrical layer stack is hardly possible. This can lead to warpage problems. If the millimeter wave substrates contain PTFE (polytetrafluouroethylene), some processing steps in the production of the PCBs are tricky. During mechanical drilling, for example, care must be taken to ensure that the PTFE is not smeared across several inner layers of the PCB. When laser drilling, care should be taken to ensure that no PTFE vapors are deposited on the PCB. In some production steps, such as the application of solder resist, it must be ensured that no substances are deposited in the millimeter wave substrate, which would lead to a change in permittivity.

The practical limit of the applicability of millimeter wave PCBs at approx. 100 GHz is caused by the tolerances when etching the metal structures. Even in etching processes with low tolerances, etching errors in the range of 10–15 μm are common. Especially when narrow slots, corners or defined thin lines occur in planar millimeter wave components, these etching tolerances are too large in relation to the wavelength at 100 GHz. Even for millimeter wave PCBs with planar components in the frequency range around 77 GHz, the tolerances must be taken into account in the design of the components in order to enable reliable production of the PCBs.

Inorganic multilayer technologies such as LTCC (low temperature co-fired ceramics) and HTCC (high temperature co-fired ceramics) are repeatedly considered in the scientific literature for applications in the millimeter wave range. However, due to the large geometric tolerances when sintering the ceramics, their use in the millimeter wave range is difficult and these technologies have so far only been able to establish themselves to a limited extent.

9.2.3 Millimeter Wave Packages

To protect the MMICs from environmental influences and make them accessible for standard surface-mounted device (SMD) soldering processes, the MMICs are usually integrated into a package. Depending on the antenna concept, the antenna is also integrated into the package, as described in Sect. 9.1. Different antenna concepts are available depending on the frequency range and production volume.

In the frequency range at 77 GHz, packages such as the eWLB(embedded wafer level ball grid array), which is basically a ball grid array, have become established due to the high production volumes. During production, the individual MMICs are glued onto an artificial wafer and then encapsulated with mold compound. After removing the wafer, the electrical connections are realized using thin-film technology. This creates the so-called redistribution layer, in which very finely structured millimeter wave lines and passive components are implemented. As the redistribution layer can be significantly larger than the MMIC, a large number of connections can be realized. The solder balls are placed on this electrical layer. Antenna structures can be realized in the redistribution layer or on the mold compound. Due to the complex manufacturing process, the eWLB package is particularly suitable for large production volumes.

The QFN (quad-flat no-leads) package, which is also used for systems in the upper millimeter wave range, is a flexible alternative, especially for smaller production volumes. Here, MMIC with antenna can be integrated directly into the package, the low-frequency signals up to approx. 40 GHz can be routed to the MMIC via the package. The MMICs are wire bonded into the package. At frequencies above 100 GHz, the antenna structure is electromagnetically coupled to the MMIC, as explained in Sect. 9.3.3. To protect the MMIC, it can be encapsulated with mold compound. However, it must be ensured that the antenna is not disturbed by the mold compound, or the mold compound must be taken into account when designing the antenna. A design example with mold compound is shown in Fig. 9.1. Using suitable mold tools, a cavity can be created in the mold compound in the area of the antenna.

In the upper millimeter wave range, new package variants made of glass are also discussed, as shown in Fig. 9.2. Glass has very low losses in the millimeter wave range, can be structured three-dimensionally and the metal structures can be realized with extremely low tolerances. However, the disadvantage of using glass is its poor thermal conductivity, which makes it difficult to cool the MMICs.

Fig. 9.1 QFN package for a radar MMIC at 160 GHz. The antenna is cut out of the mold compound to reduce the losses of the antenna. The size of the package is approximately $5 \times 5\,\text{mm}^2$

Fig. 9.2 Package made of glass for a radar-MMIC at 180 GHz. A plug connection is inserted in the center of the housing to connect a dielectric waveguide or to mount a dielectric antenna

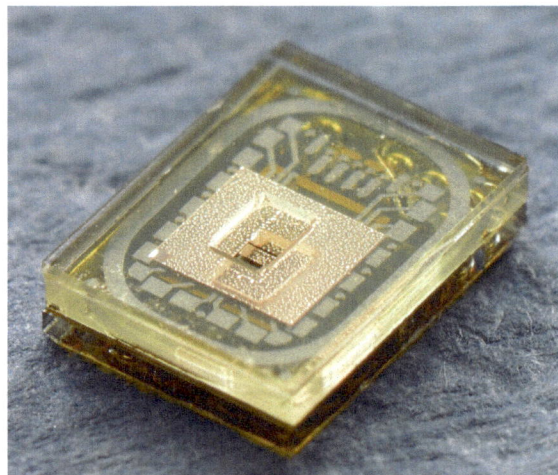

9.2.4 *Transmission Lines and Transitions*

Basically, all transmission lines that are also used at lower frequencies are used in the millimeter wave range. However, as mentioned above, the tolerance requirements for materials and geometries are significantly more challenging than at lower frequencies. In PCB technology up to approx. 100 GHz, microstrip or coplanar lines are often used. In addition, waveguides can be implemented in the PCBs as so-called substrate-integrated waveguide (SIWs), see Fig. 9.3. This conductor system is well shielded and has low losses, but is complex to implement.

9.2 Assembly and Interconnect Technology for Antennas and MMICs

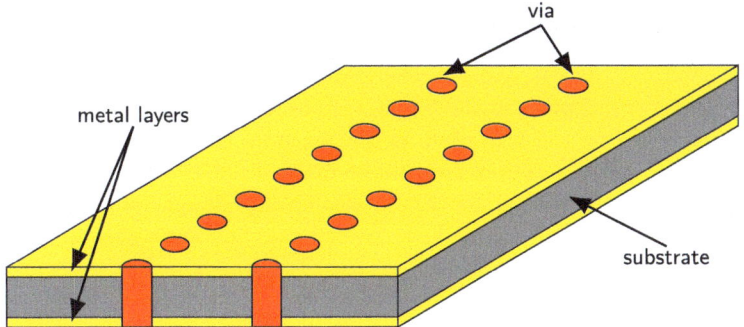

Fig. 9.3 Transmission line structure in PCB technology as a so-called SIW. The waveguide is formed by two rows of vias and two metal planes at the top and bottom

In the backend of the MMICs, essentially the same transmission line types are used as for millimeter wave PCBs. These can be used over the entire millimeter wave range because the manufacturing tolerances are orders of magnitude smaller than in PCB technologies. However, the losses of the transmission lines per unit length in the backend of MMICs are significantly higher than on a PCB, as the geometrical structures, especially the layer spacing, are much smaller and the substrate material in the bulk of the MMICs is very lossy, see Sect. 9.4. On the other hand, transmission lines in the MMIC are usually very short, thus the losses are acceptable. The losses in PCB technologies are often around 0.5–1 dB per cm of transmission line length, on an MMIC at 1 dB per mm.

Classic metallic waveguides are rarely used anymore in the upper millimeter wave range, as this technology is often too complex and expensive. Split-block technology is therefore mainly used in applications with extremely high requirements and low cost pressure. Waveguides are also used to implement transitions between MMIC and other transmission line types, or as an antenna feeding structure, see Sect. 9.3.2. New manufacturing technologies are used for this purpose, in particular 3D printing, injection moulding with metallization or multilayer sheet metal structures, in order to achieve low manufacturing costs. The dielectric waveguide has proven to be a particularly low-loss transmission line type in the millimeter wave range, with which significantly lower losses can be achieved than with classic metallic waveguides. A disadvantage of dielectric waveguides is that the field is also guided outside the conductor, which makes fixing and mounting the conductor a challenge.

Transitions between transmission line types, MMICs, antennas or any components are available in the millimeter wave range and can be found in the technical literature. The third spatial dimension is often required for the design of a broadband transition, i.e. transitions in purely planar technologies are challenging to design.

A special feature for connecting MMICs are classic wire bond transitions, e.g. a wire bond on a PCB to an MMIC. As these wire bonds are no classical millimeter wave components, they have massive parasitic effects. The wire bond acts as an inductor and radiates. In order to control the parasitic effects and take them into account in the design, the bond connections must always be identical, i.e. reproducible, and should be compensated for both in the MMIC and on the PCB. In addition, the shortest possible bond wires must be realized, whereby, for example, low parasitic inductances are achieved. For this purpose, the MMICs are usually embedded in cavities in the PCB so that bonding can take place in a plane from the MMIC to the PCB, see Fig. 9.4. Such bond connections can still be realized reliably up to a maximum of 100 GHz. A differential bond connection also enables designs at millimeter wave frequencies, as these connections have lower losses and less radiation, but above 100 GHz the manufacturing tolerances are usually too high.

9.3 Antennas

The antenna as the free space interface of the radar sensor defines the FoV and is therefore a key component for the performance of a sensor. A basic distinction must be made between radar sensors that have a constant main beam direction and those that are used for beamforming and therefore require an antenna array with many receive or transmit channels. Some frequently used antennas in the millimeter wave range are presented below, which can be used in arrays according to Chap. 6.

9.3.1 Patch Antennas

Patch antennas are particularly suitable for building antenna arrays in planar technologies such as PCB technology. A patch antenna is a $\lambda/2$-sized metal structure, the so-called patch, which is made to resonate. The patch is placed above a ground layer. A millimeter wave substrate is located between the patch layer and the ground

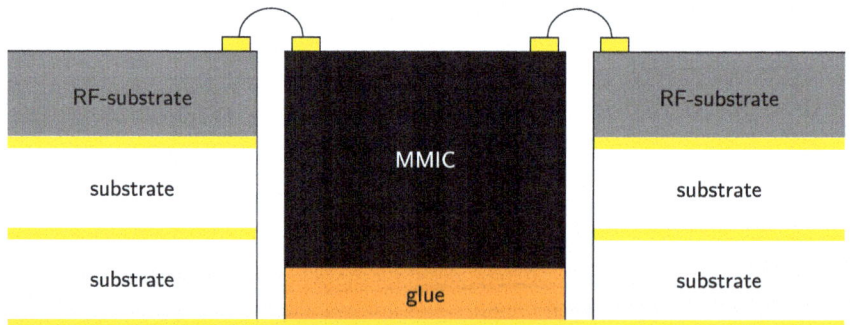

Fig. 9.4 The embedding of an MMIC in a cavity in a millimeter wave PCB leads to short bond wires. The cavity is created during the production of the PCB and then the chip is glued to a metal surface in the cavity before it is bonded

9.3 Antennas

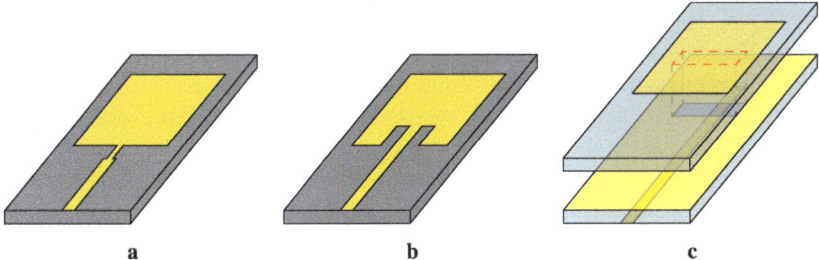

Fig. 9.5 Concepts for feeding patch antennas: **a** Direct feeding of a patch with a $\lambda/4$ transformer as a microstrip line. **b** Feeding a patch via an inset feed. **c** Aperture-coupled patches require at least two RF substrate layers that are decoupled by a ground layer with a slot

layer and the feed can be implemented in different ways, as shown in Fig. 9.5. The most common method is to feed the patches directly at the edge of the patch. Using an inset feed, in which the feed line is routed into the patch via cut-outs, the transmission line impedance and the patch can be matched to each other. The advantage of this feed method is its simplicity, as only one ground layer and one signal layer are required. Alternatively, the antenna can be fed directly from below through a coaxial structure. In the millimeter wave range, the coaxial structure is realized, for example, by through-hole plating in PCB technology or in other multilayer technologies. The widest bandwidth of patch antennas can be achieved by feeding via aperture coupling. The energy is coupled into the patch through a slot in the ground layer. The transmission line is routed under the ground layer so that the ground layer completely separates or shields the radiated field and the fields of the feeding network.

Typical relative bandwidths of patch antennas are in the range of 5–8%. This is too low for some applications in the millimeter wave range. In such cases, the bandwidth of the patch antennas can be slightly increased, e.g. by parasitically coupled structures or so-called stacked structures, i.e. several stacked patches.

Especially in the frequency range around 77 GHz, arrays of patch antennas are used in many different ways. At this frequency, the antennas including the feed network can still be manufactured using PCB technology, which enables a cost-effective antenna realization despite the use of millimeter wave substrates. However, the losses in the feed network of the antennas are generally not negligible; in fact, they limit the maximum achievable array gain. In the frequency range above 100 GHz, it is very difficult to build patch antennas using classic multilayer technologies, as these have too large tolerances.

9.3.2 Waveguide Antennas

Waveguide slot antennas are often used in the millimeter wave range. The slots in a rectangular waveguide, which deflect the current flow around the slot, lead to radiation. This enables the design of large antenna arrays with many radiating antenna elements as slots. The losses in the rectangular waveguide feeding network are often lower than it is typical of a feeding network consisting of microstrip lines, for example. The rectangular waveguides are either realized as SIW integrated in a planar multilayer structure or alternatively realized as a classical component. In the latter case, the rectangular waveguide structure is manufactured either by stacking many stamped sheets or alternatively by injection moulding components that are metallized. The signal is coupled into the waveguide either via a coupling structure on a PCB or directly from the MMIC. Especially with complex and large antenna arrays, the feeding of the antennas or the routing of the many feed lines is easier in rectangular waveguide technology than in PCB technology, as the third spatial dimension can be used anyway due to the manufacturing process. With very broadband antennas, it must be noted that feed lines in rectangular waveguide technology exhibit dispersion. If gap waveguides are used, as in Fig. 9.6, the dispersion is lower than with classical rectangular waveguides, but the losses are higher.

9.3.3 Integrated Antennas and Lenses

Beyond 100 GHz, the galvanic coupling of signals from an MMIC and their routing to the antenna becomes increasingly challenging. At these high frequencies, neither bonding wires nor other galvanic couplings allow robust designs of the transitions that can be realized in series production, as the mechanical tolerances during production are too large. However, the high frequencies make resonant antenna structures so small that they can be integrated directly into the MMIC. The antennas are realized in the metallic backend layers of the MMIC. As the backend layers in the millimeter wave range are extremely thin compared to the wavelength, the lossy silicon bulk can hardly be shielded by a continuous ground plane. As a result, the field of the

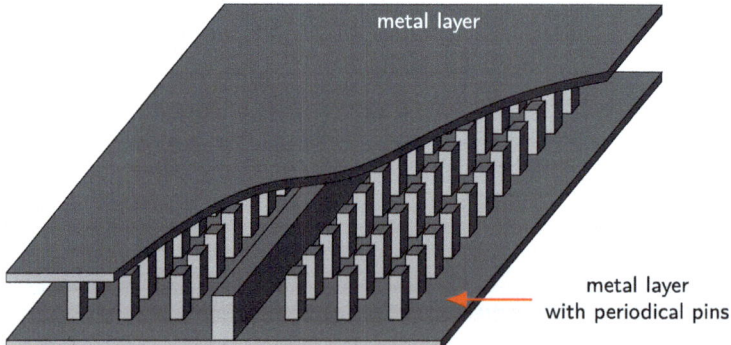

Fig. 9.6 Construction of a gap waveguide with a ridge that guides the signal. The wave is confined to the ridge and cannot propagate laterally due to the periodic pin structure

9.3 Antennas

antennas penetrates the bulk. There, the high permittivity of the silicon of $\epsilon_r \approx 11.9$ leads to the formation of a boundary layer between air and silicon, so that the field is reflected at the boundary layer and a highly lossy resonator is formed.

As a result, the efficiency of the integrated antennas, which are realized without special technologies or design measures in the backend of the MMIC, is typically limited to less than 20%. Various techniques have been developed to avoid the absorption of the electromagnetic field in the silicon bulk of the MMIC:

- Localized backside etching (LBE) allows for etching the bulk under the antenna structure in the backend layers. This results in filigree structures, but as the antenna is more or less surrounded by air after the LBE, only a small proportion of the field is absorbed in the bulk. LBE is a special manufacturing process that is not always available in industrial chip production.
- By increasing the resistivity ρ of the bulk, e.g. through doping, the losses in the bulk can be reduced. Typical values of a high resistivity are in the range of 1–10 kΩcm, in contrast to below 100 Ω cm for a normal silicon bulk. However, this approach is also a special process that is not generally available in industrial chip production.
- If external resonators, e.g. in the form of a patch or a dielectric resonator, are electromagnetically coupled directly to the backend layers of the MMIC, high antenna efficiencies can be achieved as the field hardly penetrates the lossy silicon bulk. The resonator and bulk are decoupled from each other by a metal layer in the backend. Coupling structures such as slots, meander-shaped lines or patches are used in the backend as excitation structures for the resonator. A particularly compact coupling structure is the short-circuited $\lambda/4$-patch.

Metallic patches on a low-loss superstrate such as fused silica can be used as a resonator, as shown in Fig. 9.7. Alternatively, dielectric resonators are used that radiate in the $TE_{\delta 1n}$ mode. The higher n, the higher the Q factor, which leads to better radiation efficiencies but lower bandwidths. Typical resonators have a permittivity in the range of 8–20, whereby the Q factor increases with increasing permittivity and thus the bandwidth decreases. The use of superstrates or of electromagnetically coupled external resonators also represents a special production process that must be taken into account when packaging the MMIC.

The antennas implemented in the backend are often used as primary radiator to illuminate a lens. As the chip-integrated antennas are always small compared to the wavelength, they emit a wide beam and are therefore very suitable for use with a lens. The primary radiator is placed at the focal point of the lens. Elliptical and parabolic lenses made of low-loss dielectric materials such as PTFE, HDPE (high-density polyethylene) or silicon with a high specific resistance are frequently used here.

If the primary radiator is not placed directly in the focal point of the lens, the main beam direction of the lens scans to the side ("squint") and the side lobes grow. This can be used when using several primary beams, each with its own transmit or receive direction for angle estimation and corresponds to a multi-channel monopulse

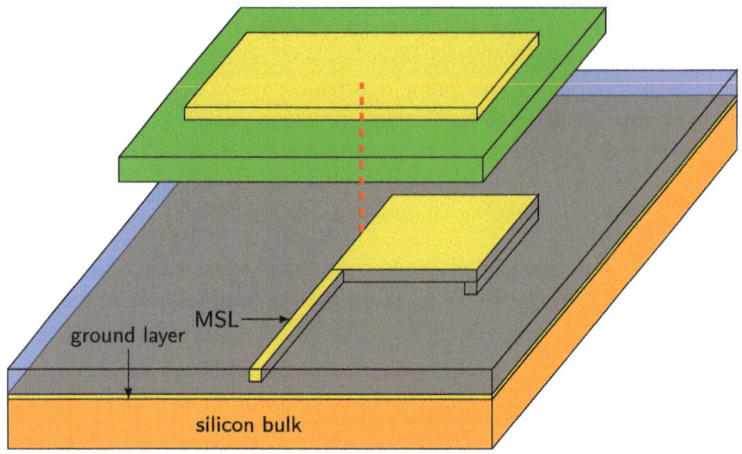

Fig. 9.7 A short-circuited $\lambda/4$ patch is integrated in the backend of the chip (*blue*), which excites a $\lambda/2$ patch on a superstrate (*green*) by electromagnetic coupling. The superstrate is positioned onto the backend layers of the MMICs

radar. However, when using bistatic MMICs, i.e. chips with separate transmit and receive primary radiators, the phenomenon of squinting leads to the transmitter and receiver having different main beam directions. This is known as parallax and severely restricts the use of bistatic MMICs at high frequencies in combination with a lens. If the parallax is very large, a target cannot be captured simultaneously by the transmit and receive pattern of the lens.

9.4 MMIC and Technology Selection

Today, MMICs are used in the millimeter wave range in various semiconductor technologies. In addition to a large number of special technologies, GaAs, Si CMOS and SiGe CMOS are mainly used. All three have different technical properties:

GaAs belongs to the group of III-V semiconductor technologies and has long been used in the millimeter wave range. GaAs allows the design of very high-performance components, as it outperforms silicon technologies in terms of the transit frequency f_T. On the other hand, GaAs is a technology that is not nearly as widespread as silicon-based technologies and therefore always has a special position. As a result, GaAs components are generally high-performance, but also expensive and are therefore less frequently used for mass products. Furthermore, logic components cannot be implemented in this technology, meaning that GaAs is only suitable for high-frequency assemblies, but not for the integration of complex systems.

SiGe CMOS By combining CMOS processes with bipolar SiGe circuits in a so-called BiCMOS process, both digital logic components and fast front-end circuits can be implemented in one technology. The bipolar transistors with low noise and higher voltages achieve ft values of several 100 GHz, which allows circuit developments in the upper millimeter wave range. This makes it the most commonly used technology in the frequency range above 100 GHz. The structure sizes available today with 130 or 55 nm are significantly larger than in modern pure CMOS processes.

Si CMOS Due to the trend towards increasing miniaturization of node sizes and the associated increase in ft in modern CMOS processes, they have become interesting for circuit development in the millimeter wave range below approx. 100 GHz. In addition to the logic components, the complete millimeter-wave front ends are integrated in CMOS. Typical CMOS implementations allow lower power losses than comparable SiGe CMOS implementations, as the CMOS transistors have lower voltages. While in the past SiGe CMOS realizations enabled higher maximum output power, larger bandwidths and better linearity, this advantage has been lost at least up to the frequency range of approx. 100 GHz today.

In many applications, especially in the frequency range below 100 GHz, in which neither Si CMOS nor SiGe CMOS is clearly technically better, the two technologies often differ in terms of cost structure. While in Si CMOS complex overall systems, which include logic components in addition to the millimeter wave front end, can be implemented on relatively small silicon areas due to the small node size, this is hardly possible in SiGe CMOS due to the much larger structure sizes. Si CMOS is therefore particularly worthwhile for applications with very high production volumes and complex overall systems. On the other hand, the development costs of such Si CMOS implementations are relatively high due to the complexity of the systems and the small structure sizes with high mask costs for semiconductor production. This is where SiGe CMOS can offer financial advantages. For small production volumes and less complex systems, the development costs in SiGe CMOS are lower, because the mask costs are not significant due to the large structure sizes and smaller wafers in contrast to small-node Si CMOS.

9.5 Signal Synthesis in Millimeter Wave Systems

The signal synthesis is a key component of many millimeter wave radar sensors. Depending on the modulation method and setup of the sensor, frequency-modulated and single tone signals must be synthesized in the millimeter wave range. For this purpose, PLL and direct digital synthesis (DDS) circuits are used, which derive the millimeter wave signal from a reference. The PLL is an analogue control loop that uses a phase-frequency detector to compare and minimize the deviation between a highly stable reference signal and the frequency divided millimeter wave signal. The output signal of the phase-frequency detector controls the oscillator, which generates the millimeter wave signal, via a lowpass filter. With a DDS, on the other hand, the frequency-modulated or single tone output signal is generated directly via a digital-to-analog conversion and is multiplied or mixed into the targeted frequency range. DDS-based signal syntheses are often more flexible, cover larger bandwidths and achieve lower phase noise than PLL systems. However, DDS approaches usually suffer from significantly larger spurious signals, higher power consumption and higher costs than PLL-based systems.

Large multiplication factors M or divider factors N, which convert the signals from the low output frequencies of the DDS or the controllable oscillator in the PLL into the millimeter wave range, are problematic in signal synthesis. These factors are squared in the phase noise power density $\mathcal{L}$

$$\mathcal{L}_{\text{mult,out}}(\delta f) = M^2 \mathcal{L}_{\text{in}}, \tag{9.2}$$

$$\mathcal{L}_{\text{div,out}}(\delta f) = \mathcal{L}_{\text{in}}/N^2. \tag{9.3}$$

In particular, the signal synthesis of frequency-modulated signals requires PLLs with large loop bandwidths, which lead to a high phase noise. If large factors M and N are used here, this leads to a further deterioration of the phase noise in the millimeter wave range. In these cases, signal synthesis architectures are used in which the frequency-modulated signal is mixed into the millimeter wave range by a single tone carrier. This prevents large multiplication factors of the frequency modulated signal. Only the single tone carrier needs to be greatly multiplied. However, the single tone carrier can be generated with low phase noise if the resonator of the PLL has a high Q factor.

Reference

1. M. Hitzler, Breitbandige hochintegrierte FMCW-Radare bei 160 GHz für industrielle Anwendungen. German, Ph.D. Dissertation, Universität Ulm, 2019. https://doi.org/10.18725/OPARU-14429

Open Access This chapter is licensed under the terms of the Creative Commons Attribution 4.0 International License (http://creativecommons.org/licenses/by/4.0/), which permits use, sharing, adaptation, distribution and reproduction in any medium or format, as long as you give appropriate credit to the original author(s) and the source, provide a link to the Creative Commons license and indicate if changes were made.

The images or other third party material in this chapter are included in the chapter's Creative Commons license, unless indicated otherwise in a credit line to the material. If material is not included in the chapter's Creative Commons license and your intended use is not permitted by statutory regulation or exceeds the permitted use, you will need to obtain permission directly from the copyright holder.

Chapter 10
Hardware Effects on System Level

The implementation of a sensor in the millimeter wave range entails not only the hardware-related challenges described in Chap. 9, but also affects the overall system design. Selected topics are therefore discussed in this chapter which illustrate the impact of hardware on system level and play a crucial role in the millimeter wave range.

10.1 Link Budget

The link budget is the power balance of a radar sensor. All power sources and drains, i.e. losses or attenuations, along the signal path are balanced, e.g. to ensure a minimum SNR or to determine the minimum required transmit power.

The transmit power is generated in the transmitter at the carrier frequency and is emitted via the antenna. However, the link budget for the transmitter not only considers the transmit power, but also all loss mechanisms such as transmission line losses or the low efficiency of the antenna. The attenuation of the radar channel essentially results from the radar Eq. (1.10), which includes the range and the RCS of the target. The antenna gains on the transmit and receive side are taken into account here. At the receiver, in turn, all losses of the circuit components or transmission lines as well as gains due to amplifiers are taken into account.

In the millimeter wave range, there are some special features in which the link budget differs compared to radar applications at lower frequencies:

- From the radar Eq. (1.10) follows that the channel attenuation in the radar channel changes with the square of the wavelength λ. This results in large channel attenuations due to the short wavelengths in the millimeter wave range. On the other hand, the antenna gain, which is included in the radar equation on both the transmit and receive side, scales with $1/\lambda^2$ for a fixed aperture size. This means that frequency scaling to higher frequencies with a constantly large aperture on the transmit and receive side leads to lower losses in the link budget. In practice, when scaling the frequency, the aperture size is usually reduced in order to build smaller sensors. If the apertures are scaled with the frequency, the link budget deteriorates with scaling to higher frequencies. Depending on the antenna type and technology, the antenna gains cannot be increased arbitrarily by scaling the antenna arrays, as the losses for feeding the many antennas in the array compensate for the increase in gain.

For linear frequency-modulated radar sensors, i.e. FMCW and chirp sequence sensors, the beat frequency is proportional to the range, see (4.21). Together with the radar Eq. (1.10) follows that the power of the beat signal decreases with $1/R^4$ over the beat frequency. This can be compensated for by a highpass filter in the baseband, which significantly relaxes the dynamic range requirements on the ADC. This is a major advantage in the hardware design of linear frequency-modulated radar sensors compared to all other modulation methods.

- Due to the high frequency, generating high output powers is challenging in terms of technology and circuitry. Typical maximum transmit power in the range of 77 GHz lies between 10–15 dbm with SiGe CMOS and Si CMOS circuits. Significantly higher transmit powers can be achieved, particularly through the use of power amplifiers in III–V semiconductor technologies. If the sensors are not limited in terms of SNR by thermal noise but by phase noise or clutter, increasing the transmit power does not improve the performance of the sensors. For this reason, the low maximum transmit power is not a serious limitation in many typical short-range applications.

It should also be noted here that the so-called equivalent isotropic radiated power (EIRP) is given by

$$P_{\text{EIRP}} = G_{\text{tx}} P_{\text{tx,max}}. \qquad (10.1)$$

This value originates from the context of frequency regulation and is proportional to the maximum power density that a radar system may excite. Due to the large antenna gains in the millimeter wave range, the EIRP is usually significantly greater than the transmit power.

- The large bandwidths in the millimeter wave range and the high number of channels of imaging radars lead to very large integration gains in many applications, which significantly improve the SNR, as explained in Sect. 8.3. Without the integration gains, the SNR is negative in many applications in the millimeter wave range. Only the high integration gains allow to obtain an SNR greater than zero decibels.

10.2 Leakage and Short-Range Behavior

Leakage in the context of radar sensors refers to the unwanted or parasitic coupling of the transmit signal within the radar sensor into the receiver. This parasitically coupled signal often has a level many orders of magnitude larger than the desired received signal from the target and superimposes on it. Mixing the received signal and the

10.2 Leakage and Short-Range Behavior

superimposed leakage signal with the transmit signal results in a high signal level at or near zero frequency, which is problematic in terms of circuitry and significantly restricts the short-range behavior of, e.g., FMCW sensors. The parasitic coupling is usually not caused by a single coupling path, but by a large number of paths. Reflections on the radome, reflections due to antenna matching, coupling paths on the PCB or in the MMIC contribute to leakage. Each path can have a different length and attenuation, which ultimately leads to the leakage signal varying greatly in amplitude and phase over the frequency when all signals are superimposed, i.e. the signal is randomly modulated. Basically, this corresponds to a fading effect as occurs in multipath propagation. The longer the individual coupling paths are, the more frequency-dependent their superposition or modulation becomes. When mixing into the baseband, this in turn leads not only to a constant signal at zero frequency, but also to the beat signal spectrum being filled with interference signals at low frequencies, as shown in Fig. 10.1. The first two to five range cells are often affected by this, which leads to poor short-range behavior of the radar sensors.

Furthermore, the strong leakage signal leads to high demands on the receiver circuitry, especially on the input stage. In the classic design, an LNA is used as the first component in order to keep the chain noise figure low. The combination of the requirements for the LNAs of low noise figure, high gain and a large dynamic range caused by the leakage signal often cannot be met in the millimeter wave range. For

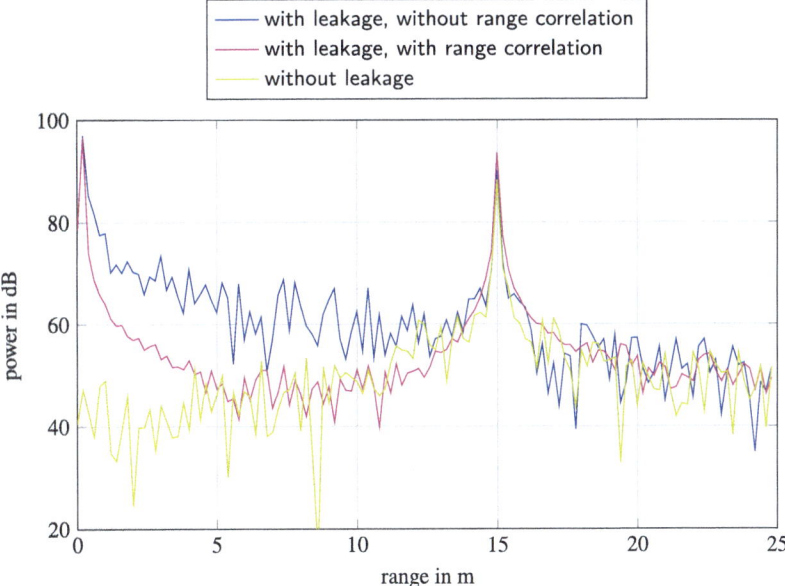

Fig. 10.1 Spectrum of a simulated beat signal for a target at 15 m with and without leakage effect. For small ranges, leakage causes a significant increase in the noise floor. The range correlation reduces the leakage effect, but does not completely eliminate it

this reason, often no LNAs or LNAs with a very low gain are used, which results in a relatively high receiver chain noise figure. Here, circuits or LNAs in III–V semiconductor technology are superior to SiGe CMOS and Si CMOS circuits.

Countermeasures that improve the close-range behavior of the sensors generally follow two strategies. Firstly, the number of coupling paths can be reduced. Strong coupling paths and paths of great length should be reduced, as these have a particularly negative effect on leakage. These include the paths caused by reflections on the radome and by antenna matching. For example, it is important to ensure that the distance between the antenna and radome is as short as possible in order to minimize the path length. Secondly, the dominant coupling paths, should have a constant amplitude over the frequency, as far as this can be influenced and they should not be strongly frequency-dependent. This can usually be achieved, at least when matching the antenna. The frequency-selective behavior of the leakage signal can be reduced through constant matching.

10.3 Phase Noise

As explained in Sect. 9.5, phase noise plays a major role in signal synthesis in the millimeter wave range. Large frequency multiplication factors should be avoided as they increase the phase noise, see (9.2). If the phase noise is too high, strong targets whose signal peak is affected by the phase noise may increase the noise floor in the spectrum of the beat signal. This can mask weak targets.

The transmit signal with phase noise is reflected by the target and received again by the receiver with a time delay. If the transmitter and receiver operate incoherently, the phase noise of the transmitter adds to the phase noise of the receiver, so that the beat signal is subject to the addition of both phase noise components.

In the millimeter wave range, however, practically only coherent sensors are used in which the phase noise on the transmitter and receiver side are not independent of each other. This is where the so-called range correlation effect occurs, particularly in short-range applications, which is described in the following using the example of an FMCW sensor.

The behavior of a radar can be described in simplified terms by a transfer function. The transfer function is made up of the transfer function of the transmit side up to the target, as well as that of the receive side, i.e. the radar channel from the target to the receiver and the receiver itself. If the transmission-side transfer function is normalized to one, the reception-side transfer function is identical, but it must represent the time shift τ given by the propagation time in the channel. Mixing the transmit signal with the receive signal corresponds to a subtraction in the frequency domain, resulting in the following simplified radar transfer function

$$H(\omega) = 1 - e^{-j\omega\tau}, \tag{10.2}$$

10.3 Phase Noise

where τ is the transit time of the signal in the radar channel. If this transfer function is to be applied to the phase noise of the LO signal, i.e. the spectral power density of the phase fluctuation, it must be squared:

$$|H(\omega)|^2 = 2(1 - \cos(\omega\tau)). \tag{10.3}$$

With the phase noise $S_{tx}(\omega)$ of the LO signal or the transmit signal, this results in

$$S_{beat}(\omega) = 2 S_{tx}(\omega)(1 - \cos(\omega\tau)) \tag{10.4}$$

for the phase noise of the beat signal. In case of extreme close ranges, in which $\tau \approx 0$ applies, the phase noise of the LO has practically no influence on the beat signal. This is due to the fact that the phase noise on the transmit and receive side are strongly correlated due to the short channel delay and therefore almost completely cancel each other out when mixing into the baseband. However, the longer the channel delay τ, the stronger the decorrelation between the transmit and receive phase noise, i.e. the larger the target range, the stronger the effect of the phase noise.

In addition to the received signal, the leakage signal occurs in the receiver as described in Sect. 10.2. This is subject to the transmit-side phase noise and is mixed into the baseband by the mixer. Due to the very short propagation time of the leakage signal, the range correlation is very large, which leads to a strong suppression of the leakage phase noise. However, the levels occurring here are so high compared to a received signal from a target that sensors in the millimeter wave range can be limited by the phase noise in the SNR despite the range correlation, as shown in Fig. 10.1. In these cases, the SNR cannot be improved by increasing the transmit power, as the level of the phase noise is increased with the transmit signal level. Suitable sensor architectures that are optimized for low phase noise in the signal synthesis and strong attenuation of the leakage signal can help here.

Open Access This chapter is licensed under the terms of the Creative Commons Attribution 4.0 International License (http://creativecommons.org/licenses/by/4.0/), which permits use, sharing, adaptation, distribution and reproduction in any medium or format, as long as you give appropriate credit to the original author(s) and the source, provide a link to the Creative Commons license and indicate if changes were made.

The images or other third party material in this chapter are included in the chapter's Creative Commons license, unless indicated otherwise in a credit line to the material. If material is not included in the chapter's Creative Commons license and your intended use is not permitted by statutory regulation or exceeds the permitted use, you will need to obtain permission directly from the copyright holder.

Part IV
Advanced Radar Topics

Chapter 11
Radar-Based Grid Maps

In this chapter, radar-based mapping of the environment for robotic applications or for automated driving is presented. The previous chapters outlined the basic structure of millimeter-wave radar sensors and the associated signal processing methods for detecting, localizing and determining the velocity of targets. This enables precise spatial localization of each detected target relative to the radar sensor. Methods for transforming the data collected relative to the sensor into a global map and for localizing the sensor in this map are described in this and the next chapter.

The generation of a map of the environment is the basis for a wide range of applications in the field of robotics and autonomous systems, including the planning of driving trajectories, obstacle avoidance, or finding parking spaces [1]. Grid maps have been known since the 1980s [2] and provide a simple and effective method for storing, processing, merging and presenting information about the environment. The two-dimensional visualization of a grid map from a bird's eye view allows an intuitive interpretation of the maps without in-depth technical knowledge. The efficient generation of grid maps makes them a fundamental basis for a wide range of applications.

Grid maps segment a map into a uniform grid, either in two-dimensional or three-dimensional space, in order to create a corresponding image of the environment. In the field of autonomous driving, grid maps are used, for example, to map parking lots, intersections or road traffic, as illustrated in Fig. 11.1. In front of the red vehicle are various objects and obstacles, that are detected by the radar sensor built into the vehicle and then localized as described in Chap. 6. By means of grid maps, the object information obtained from the radar data is transferred into a precise map of the surrounding area.

An ideal grid map contains only three well-defined states:

- free cell
- occupied cell
- unknown cell information.

© The Author(s) 2025
C. Waldschmidt et al., *Millimeter Wave Radar*,
https://doi.org/10.1007/978-3-031-89118-2_11

Fig. 11.1 Exemplary visualization of a street scenario (*top*) with the corresponding grid map (*bottom*). The grid map is divided into three states: free (*blue*), occupied (*yellow*), and unkonwn information (*turquoise*)

In Fig. 11.1, the trees, the person and the vehicle are clearly mapped as occupied cells (*yellow*) onto the grid map at the bottom, while most of the other area is depicted as free (*blue*). Uncertain cell information (*turquoise*) arises outside the radar sensor's field of view, as well as due to occlusion effects or volatile detections. Although a grid map could theoretically store information from only one measurement, in practice grid maps are used to fuse measurement data from multiple measurements. Fusing many measurements provides a more robust representation of the environment. This results from the gain of integrating many measurements and from capturing objects to be detected from different viewing angles (as soon as the radar is moving), thus reducing occlusion effects.

The grid map of the environment as sketched in Fig. 11.1 is described in the following by means of the matrix **U**. This matrix subdivides the environment into N_ϱ cells, where each cell ϱ is identified by a unique index n_ϱ. For ease of reading, the index n_ϱ will be neglected in the following in the context of a cells-based description, unless a unique designation is necessary. Depending on the respective algorithm, the individual cells of the grid map represent either receive signal power, as described in Sect. 11.1, or occupation probabilities, as explained in Sect. 11.2. Each cell ϱ of the mapped environment is associated with a specific position $P_\varrho = [x_\varrho, y_\varrho]$ in Cartesian coordinates.

The basic structure of an algorithm for creating grid maps consists of three steps, as visualized in Fig. 11.2 for a sensor detecting two targets (*red crosses*). The same steps are carried out for each radar measurement and, in the case of multiple sensors, also for each sensor:

1. Creating a local map
2. Coordinate transformation
3. Map update of the global map.

In the illustration in Fig. 11.2, the local map in the sensor coordinate system is represented by the *gray grid*, while the global map, e.g. a parking lot, is represented by the *black grid*. The first step is to create a local map in the sensor coordinate system that represents the immediate surroundings of the vehicle at the current time. This local map must then be transformed into the coordinate system of the global map. This transformation is based on the vehicle's ($_F$) pose information $\mathbf{P}_F^g$ in the global coordinate system (^{g}) by means of a rotation and subsequent translation. The information about the environment obtained in this way is integrated into the existing global map by means of a map update. This integration process is carried out by fusing the local and global maps. If this process is continued iteratively, the global map contains the information from all previous measurements, thus enabling a precise representation of the environment. This basic structure is identical for the two most common grid mapping methods, amplitude grid map (AGM) and probabilistic occupancy grid map (OGM). The main difference between the two methods lies in the type of information represented and in the procedure for creating the local map and the map update.

In Sects. 11.1 and 11.2, AGMs and probabilistic OGMs are presented, followed by a comparison in Sect. 11.3. Subsequently, the simultaneous localization and mapping (SLAM) method is presented in Sect. 11.4, which enables self-localization within the global map.

11.1 Amplitude Grid Map (AGM)

One of the simplest and computationally most efficient implementation options for grid maps are so-called amplitude grid maps (AGMs). These are based on an evaluation and representation of the targets in the grid map using amplitudes. The basic procedure of the amplitude-based grid map algorithm comprises the three general steps presented in the introduction of this chapter.

In the first step, a local grid map is created in the specific sensor coordinate system and contains all targets of the measurement currently being analyzed. The target information is determined using the signal processing methods from Chaps. 5 and 7, i.e. using a two-dimensional FFT and a corresponding consecutive angle estimation. Through this, the range R_q, the azimuth angle ψ_q in relation to the radar sensor, and the amplitude A_q are obtained for each of the Q detected targets. These targets are depicted as *red crosses* in the sensor coordinate system in Fig. 11.2.

Due to measurement uncertainties, non-ideal localization and a discrete target spectrum, the measured values or, respectively, the target position estimates include errors. These errors are described by means of a two-dimensional Gaussian distribution according to

Fig. 11.2 Visualization of the three basic processing steps of grid maps. **Step 1** the two exemplary detected targets are added to the local map with their uncertainty. **Step 2** involves transforming from the local sensor coordinate system to the global coordinate system. **Step 3** involves updating the map and thus fusing the current measurement with all previous measurements

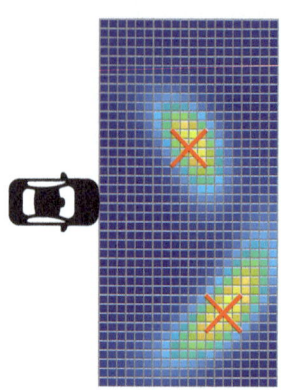

step 1: generate local map

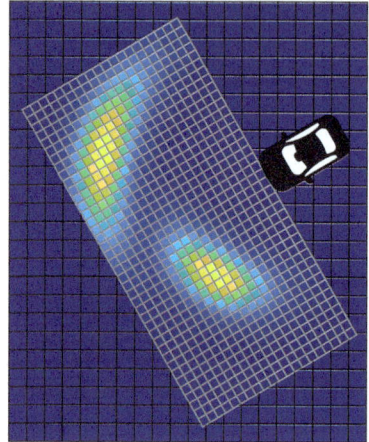

step 2: transform local map to global map

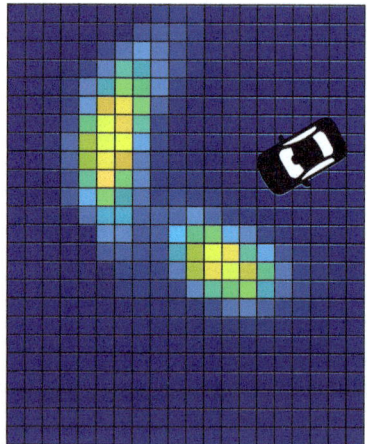

step 3: fuse local and global map

11.1 Amplitude Grid Map (AGM)

$$\chi_{\mathbf{P}_\varrho}\left(\mathbf{P}_\varrho^{s,p}\right) = \frac{1}{\sqrt{(2\pi)^2 \det(\mathbf{\Sigma})}} e^{-\frac{1}{2}(\mathbf{P}_\varrho^{s,p}-\mu_q)^T \mathbf{\Sigma}^{-1}(\mathbf{P}_\varrho^{s,p}-\mu_q)}, \quad (11.1)$$

where $\mathbf{P}_\varrho^{s,p}$ describes the position of the pixel $n_\%$ to be evaluated in polar coordinates (^{p}) in the sensor coordinate system (^{s}). The expected value vector μ_q contains the measured range and azimuth angle at which the target is detected, and $\mathbf{\Sigma}$ is the covariance matrix, which is proportional to the standard deviations from (3.25). The representation of the q-th target in cell ϱ_{n_v} thus corresponds to a two-dimensional Gaussian distribution around the maximum amplitude of the received power

$$A(\varrho^s) = A_q \cdot \frac{\chi_{\mathbf{P}_\varrho}\left(\mathbf{P}_\varrho^{s,p}\right)}{\chi_{\mathbf{P}_\varrho}(\mu_q)}, \quad (11.2)$$

where $A(\varrho^s)$ describes the received power of the targets in relation to the sensor coordinate system. For reasons of computational efficiency, it can be useful to neglect the power from a range of more than $3\sigma_R$ and 3σ and thus consider it to be zero.

After creating the local map $\mathbf{U}^s$ in the sensor coordinate system, a transformation into the global coordinate system is required. For this, both the global vehicle pose (position and orientation) $\mathbf{P}_F^g$ and the position of the sensor ($_S$) $\mathbf{P}_S^f$ in the vehicle coordinate system (^{f}) must be known. The vehicle pose is usually determined using global navigation satellite system (GNSS) or a SLAM algorithm, as described in Sect. 11.4. The transformation is carried out in two steps: first a rotation and translation into the vehicle coordinate system is carried out, followed by a second rotation and translation into the global coordinate system. The result of this transformation is shown in Fig. 11.2 through the superimposition of the global map (*black grid*) and the local map (*gray grid*).

The last step is the map update, which allows a fusion of all previous measurements with the current measurement or a fusion of the already existing global map with the currently generated map. For AGMs, this map update is realized by summation of corresponding cells:

$$A\left(\varrho^g \mid \mathbf{\Lambda}_{1:t}, \mathbf{P}_{F_{1:t}}^g\right) = A\left(\varrho^g \mid \mathbf{\Lambda}_t, \mathbf{P}_{F_t}^g\right) + A\left(\varrho^g \mid \mathbf{\Lambda}_{1:t-1}, \mathbf{P}_{F_{1:t-1}}^g\right) \quad (11.3)$$

where $\mathbf{\Lambda}$ describes the radar measurement in general. The current measurement at time t and thus the detected environment, which is transformed into the global coordinate system, is represented by the first term $A(\varrho^g \mid \mathbf{\Lambda}_t, \mathbf{P}_{F_t}^g)$. The second term $A(\varrho^g \mid \mathbf{\Lambda}_{1:t-1}, \mathbf{P}_{F_{1:t-1}}^g)$ describes the amplitude of the global map at time $t-1$. In each cell of the final map, the accumulated received power over the entire measurement period is visualized, as exemplarily shown in Fig. 11.2.

The entire calculation and representation in an AGM are based exclusively on amplitudes. As a result, targets with a large RCS and a short range to the detecting radar sensor are represented by a high amplitude $A(\mathbf{U})$ in the map if they are detected in many frames. The dynamics of the resulting map are therefore not controllable and can potentially cause a loss of information, especially for weak or distant targets. Another disadvantage lies in the map update, which is based exclusively on a summation. False targets therefore cannot be removed from the map, which leads to the permanent storage of false information in the map. These two disadvantages, can be avoided by using probabilistic OGMs, which, however, are significantly more complex to create.

11.2 Probabilistic Occupancy Grid Map (OGM)

In contrast to AGMs, probabilistic occupancy grid maps (OGMs) enable a probabilistic calculation and representation of the environment. With OGMs, the cells do not represent amplitudes, but occupancy probabilities $p(\mathbf{U})$, which allows a direct state estimate for each cell. The goal is to estimate the occupancy probability $p(\varrho^g \mid \mathbf{\Lambda}_{1:t}, \mathbf{P}^g_{F_{1:t}})$ of cell ϱ at time t based on all measurements $\mathbf{\Lambda}_{1:t}$ and all pose information $\mathbf{P}^g_{F_{1:t}}$ acquired up to the current time t. The occupancy probability p can be considered as the existence probability of an object in the cell. For $p = 0.5$, no information is available about the corresponding cell; $p > 0.5$ indicates targets and $p < 0.5$ models the free space.

The basic structure (local map, transformation, and map update) of OGMs is identical to the AGM and to the general structure from Fig. 11.2. In the first step, a local map is created based on the target lists. This map is subsequently transformed into the global coordinate system and fused with the existing map. However, both the creation of the local map and the map update differ significantly from those of amplitude-based grid maps. This is described in the following, taking into account [3–5].

11.2.1 Local Map

The local map for probabilistic OGMs represents the occupancy probability of each cell. To model this occupancy probability, first the free space is modeled using the occupancy probability $p_{FS}(\varrho)$ and afterwards the targets are modeled using the occupancy probability $p_T(\varrho)$. The local map is obtained by merging these two models.

11.2.1.1 Free Space Model

The free space model describes the reliability of the radar sensor. If the radar sensor does not detect a target in an area of the map, the free space modeling is used to determine the probability that there is actually a target-free area or space in front of the radar; hence the name free space. The probability of detection decreases with

11.2 Probabilistic Occupancy Grid Map (OGM)

increasing range for radar sensors due to the R^4 dependency, according to (1.10), as well as at the edges of the FoV. Thus, the probability of the free space ($_{FS}$) can be determined based on these influencing factors using

$$p_{FS}(\varrho) = \frac{0.5 - c_1}{2} \cdot \left(\frac{R_\varrho^4}{R_{ua}^4} + \left(1 - \cos\left(\frac{\pi}{2} \frac{\psi_\varrho}{\mathcal{F}_\psi} \right) \right) \right) + c_1 , \qquad (11.4)$$

whereby the angular dependency is approximated by the cosine. The variable R_{ua} indicates the maximum detectable range of the radar sensor, while $\mathcal{F}_\psi$ describes the FoV, i.e. the maximum beamwidth of the utilized antennas in the azimuth plane. The minimum of the free space model is determined by the variable c_1, which has a significant influence on the de-integration process and thus on the suppression of false targets. A possible free space model is illustrated in Fig. 11.3a. In the free space model, the probability of occupancy lies in the interval $p_{FS} \in [0; 0.5]$, because $p < 0.5$ represents free space.

11.2.1.2 Target Model

The second step of the modeling involves the probabilistic representation of the targets. Since the target objects were extracted directly from the R-v matrix, the probability must lie in the interval $p_T \in [0.5; 1]$. In contrast to the AGM, in which the targets are mapped exclusively based on their signal power, the probabilistic OGM incorporates the existence probabilities of the targets into the map. The likelihood of the existence of a target depends largely on the SNR of the target and is modeled accordingly with

$$p_{SNR} = 1 - 0.5 \cdot e^{-0.05 \cdot SNR} . \qquad (11.5)$$

The probability p_{SNR} corresponds to the probability of detecting the target for the estimated range and estimated azimuth angle μ_q. The factor 0.05 corresponds to a target probability of 0.6 for an SNR of 10 dB. The measurement uncertainty of the target position is modeled with a two-dimensional Gaussian distribution according to (11.1). With the appropriate normalization to the maximum probability p_{SNR}, the following occupancy probability results for each cell:

$$p_T(\varrho) = \begin{cases} 0 & , \|\mathbf{P}_\varrho^{s,p} - \boldsymbol{\mu}_q\| > \mathrm{Tr}\left(3\sqrt{\Sigma}\right) \\ \frac{\chi_{\mathbf{P}_\varrho}(\mathbf{P}_\varrho^{s,p})}{\chi_{\mathbf{P}_\varrho}(\boldsymbol{\mu}_q)} \cdot (p_{SNR} - 0.5) + 0.5 & , \text{otherwise} \end{cases} . \qquad (11.6)$$

For reasons of computational efficiency, it is reasonable to neglect the probability for deviations greater than $3\sigma_R$ and 3σ and thus consider these to be zero. According to Sect. 11.1, the covariance matrix Σ is proportional to the standard deviations. A probability distribution is determined for each of the Q detected targets. This

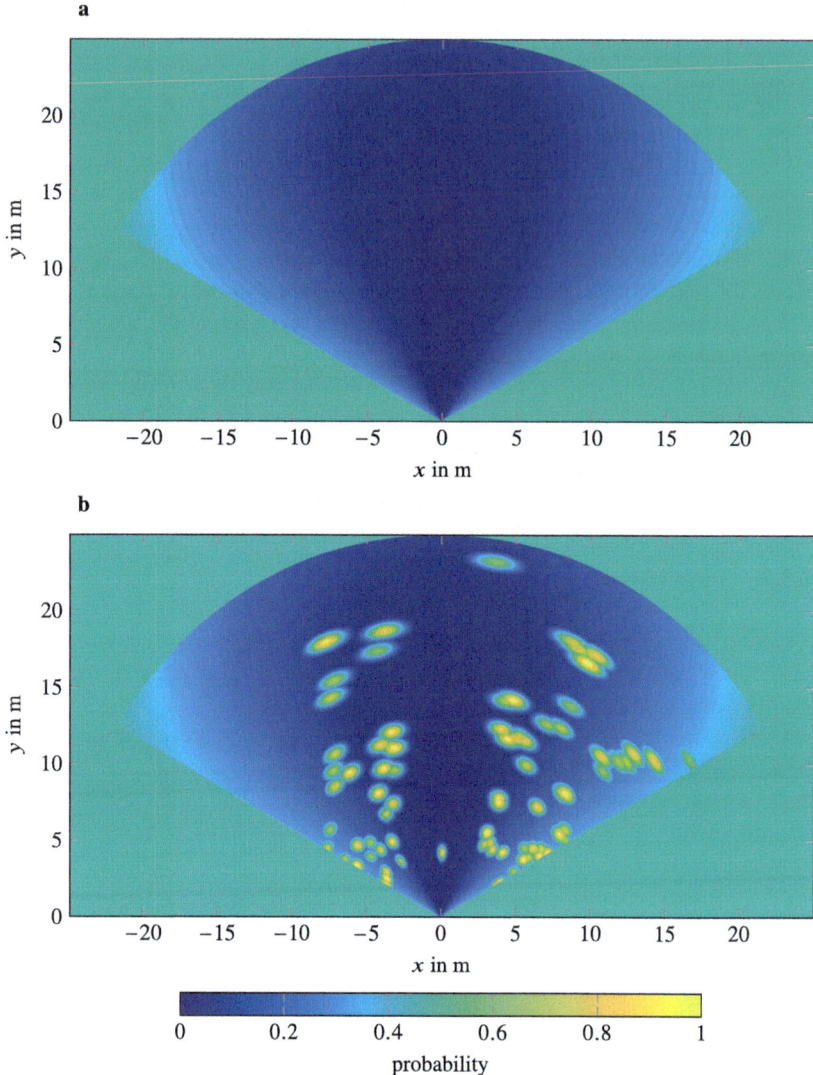

Fig. 11.3 Illustration **a** of the free space model which represents the uncertainty of the radar sensor in the case of a measurement without obstacles. Subsequently, the detected targets and corresponding measurement uncertainties are added to the map of the free space model to obtain the final local map **b**

occupancy probability is subsequently fused with the free space model according to

$$p\left(\varrho^g \mid \mathbf{\Lambda}_t, \mathbf{P}_{F_t}^g\right) = \max\left\{p_T(\varrho), p_{FS}(\varrho)\right\}. \tag{11.7}$$

The result of this fusion is illustrated in Fig. 11.3b. The free space model remains unchanged and the detected targets are inserted into the map with their individual target probabilities. This process leads to the generation of the local map. According to Fig. 11.2, the transformation into the global coordinate system and the map update follows.

11.2.2 Map Update

The map update, i.e. the fusion of the current measurements $\mathbf{\Lambda}_{1:t}$ at time t with the previous measurements $\mathbf{\Lambda}_{1:t-1}$, involves a probabilistic process for OGMs. A simple summation, as practiced in AGMs, is not useful in this case because occupancy probabilities cannot be summed. According to the principles from [5], the map update is carried out using probabilistic filter methods or Bayesian methods which enable a weighted fusion of new information:

$$p\left(\varrho^g \mid \mathbf{\Lambda}_{1:t}, \mathbf{P}^g_{F_{1:t}}\right) = \frac{1}{\left(1 + p_A^{-1} \cdot p_B^{-1} \cdot p_C^{-1}\right)} \quad (11.8)$$

where

$$p_A = \frac{p\left(\varrho^g \mid \mathbf{\Lambda}_t, \mathbf{P}^g_{F_t}\right)}{1 - p\left(\varrho^g \mid \mathbf{\Lambda}_t, \mathbf{P}^g_{F_t}\right)} \quad (11.9)$$

$$p_B = \frac{p\left(\varrho^g \mid \mathbf{\Lambda}_{1:t-1}, \mathbf{P}^g_{F_{1:t-1}}\right)}{1 - p\left(\varrho^g \mid \mathbf{\Lambda}_{1:t-1}, \mathbf{P}^g_{F_{1:t-1}}\right)} \quad (11.10)$$

$$p_C = \frac{1 - p(\varrho)}{p(\varrho)}. \quad (11.11)$$

The update step contains three parts: The information from the current measurement p_B is merged with the knowledge from all previous measurements p_A, and thus with the global map, taking into account the general a priori information of the measurement uncertainty p_C.

11.3 Comparison of AGM and OGM

Although the basic principles of data processing are similar for AGMs and OGMs, they represent the environment in different ways. While both the calculation and the representation are based on amplitudes for the AGM, they are based on probabilities for the OGM. A comparison of both representations, based on identical raw data, is shown in Fig. 11.5 for a parking lot drive. For this, a 77GHz automotive radar sensor was used, which is mounted on the side of a vehicle. The trajectory of the vehicle ranges from to coordinates (50 m, 0 m) to (120 m, 0 m) in Fig. 11.5, with the radar sensor facing into the negative y-direction. The measurement scenario is shown in Fig. 11.4.

The AGM provides a rough representation of the environment. In Fig. 11.5a, two vehicles can be observed at $x = 70$ m and four more at $x = 110$ m. Trees and street lamps are also clearly visible. Due to the cumulative update step of the AGM, the amplitude of a cell increases continuously over time, which leads to an accumulation of erroneous information and a certain blurring of the overall picture. Another peculiarity is the range dependency, whereby targets in close proximity to the trajectory have a higher amplitude than distant targets, which makes detection in more distant areas and a distinction between occupied and free cells more difficult.

The blurring and range dependency are effectively reduced by using OGMs, as shown in Fig. 11.5b. Although using identical radar parameters, the result is a much sharper representation of the environment. Vehicles show clear contours and are well distinguishable from each other. Even the trees show finer details, not just a rough point cloud. Another advantage of OGMs is in the immediate occupancy estimation of each cell. Without the need for an additional CFAR step, passable areas are directly recognizable (in Fig. 11.5b the *blue areas*). Areas with an occupancy probability of 50% indicate that the radar does not provide sufficient information about these areas for a precise state estimate. This results both from occlusion effects behind objects (for example the trees in Fig. 11.5) and from the limited range of the radar sensor (here: approximately 20 m). This relevant information about the uncertainty of certain cells is not included in AGMs, which causes free space and uncertainty to be displayed identically.

Fig. 11.4 Photo of the measurement environment with several parked cars, hedges and trees, fences, streets and gravel parking lots. The *blue line* indicates the approximate course of the trajectory

11.3 Comparison of AGM and OGM

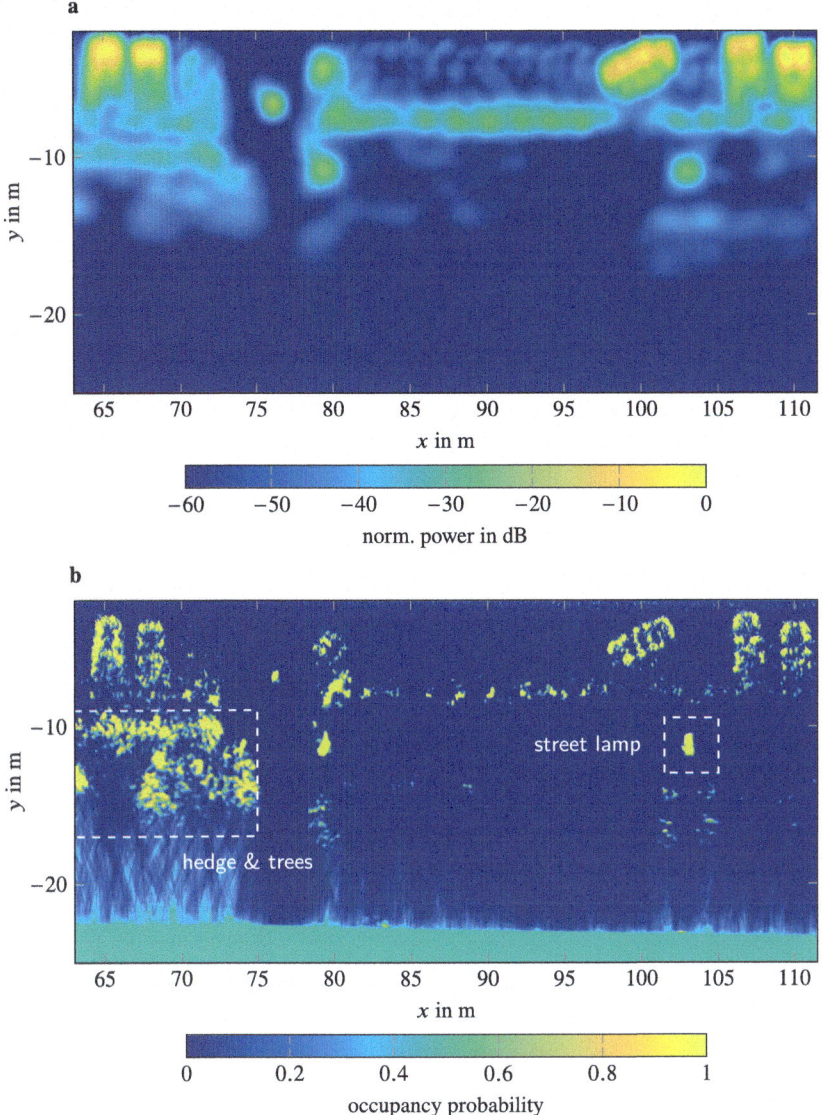

Fig. 11.5 Visualization and comparison of **a** an AGM and **b** an OGM for the scene in Fig. 11.4. Prominent landmarks such as hedges, trees and streetlamps are marked

The advantages of OGMs clearly prevail compared to AGMs, despite the more extensive processing effort. Specifically in terms of accuracy, robustness, and state estimation, more accurate environment mappings can be obtained with OGMs using identical radar hardware.

11.4 Simultaneous Localization and Mapping (SLAM)

For both AGMs and OGMs, the pose information of the vehicle in the global coordinate system must be known. Typically, the position estimate based on GNSS is used for this purpose. Conventional external localization systems, such as GNSS, reach their limits in many scenarios when the connection to the satellite is impaired, such as in tunnels, buildings or within cities. This is where the so-called SLAM method comes in as a promising localization alternative, since it relies exclusively on the vehicle's own radar data. SLAM is an innovative method by which mobile robots and autonomous vehicles can determine their position in real time while simultaneously creating a map of their unknown environment. In contrast to external localization systems, SLAM-based localization enables reliable self-localization of the vehicle. The main advantages of SLAM compared to GNSS are:

- *Independence from external signals*: SLAM is based on the internal radar sensors of the vehicle and is therefore not dependent on external signal sources such as satellites. This makes SLAM particularly suitable for environments where GNSS signals are weak or unavailable, such as in tunnels, indoors or in urban environments.
- *Independence from temporal synchronization*: For optimal localization, the global position of the vehicle must be known for every measurement. While GNSS-based localization requires the position data to be time-synchronized with the radar data, time synchronization is inherently guaranteed with SLAM.

In the following, two different approaches are presented that use SLAM methods to solve the problem of mapping and localization being dependent on one another. These approaches enable simultaneous localization and mapping in real time, making SLAM a promising technology for autonomous driving and robot-based applications.

11.4.1 Dead Reckoning Based on Ego-Motion Estimation

One of the simplest methods for self-localization is dead reckoning. In this method, the velocity information is integrated piecewise to obtain an estimate of the trajectory. Therefore, the ego-motion of the vehicle, as shown in Fig. 11.6, must be known.

The vehicle velocity vector generally consists of the two velocity components v_x and v_y, as well as the yaw rate v_ω, which are all applied at the center of the rear axle. These three velocity components can be determined, for example, using wheel speed sensors. However, these sensors are prone to errors due to slippage or skidding. One way to overcome these measurement inaccuracies is to estimate the ego-motion based on the vehicle's radar sensors. This enables an accurate and robust ego-motion estimate for the degrees of freedom v_x, v_y, and v_ω, based on the extracted angle and velocity information. For this purpose, the following model is set up, which uses

11.4 Simultaneous Localization and Mapping (SLAM)

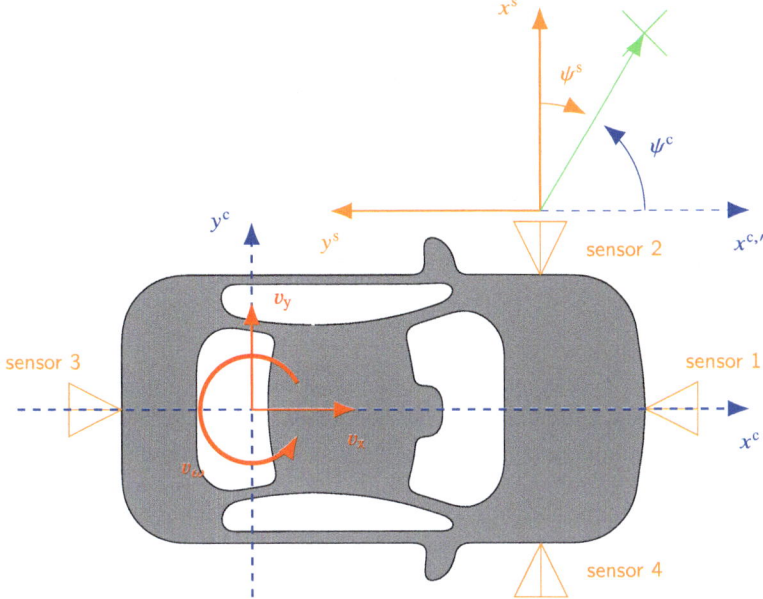

Fig. 11.6 Representation of the vehicle coordinate system (*blue*), the sensor coordinate system (*orange*), the vectorial vehicle velocities (*red*), and a target (*green*). The coordinate systems can be transformed into each other via the geometric relationships

the relationship between the vehicle's own velocity vector $\mathbf{v}_c = [v_\omega, v_x, v_y]^T$ and the measured target information (angle ψ and velocity v_r):

$$\underbrace{\begin{bmatrix} -v_{r_{1,n_s}} \\ -v_{r_{2,n_s}} \\ \vdots \\ -v_{r_{Q,n_s}} \end{bmatrix}}_{\mathbf{v}_{r_{n_s}}} = \underbrace{\begin{bmatrix} \cos(\psi^c_{1,n_s}) & \sin(\psi^c_{1,n_s}) \\ \cos(\psi^c_{2,n_s}) & \sin(\psi^c_{2,n_s}) \\ \vdots \\ \cos(\psi^c_{Q,n_s}) & \sin(\psi^c_{Q,n_s}) \end{bmatrix}}_{\psi_{n_s}} \cdot \begin{bmatrix} -y^c_{n_s} & 1 & 0 \\ x^c_{n_s} & 0 & 1 \end{bmatrix} \cdot \underbrace{\begin{bmatrix} v_\omega \\ v_x \\ v_y \end{bmatrix}}_{\mathbf{v}_c} . \quad (11.12)$$

Here, $x^c_{n_s}$ and $y^c_{n_s}$ denote the sensor positions and ψ^c_{q,n_s} the measured angle of arrival of the target signals in relation to the vehicle coordinate system obtained at sensor n_s. These must be transformed from the sensor coordinate system ψ^s_{q,n_s} to the vehicle coordinate system according to Fig. 11.6 through

$$\psi^c_{q,n_s} = \psi^s_{q,n_s} + \varphi^c_{n_s}, \quad (11.13)$$

where $\varphi^c_{n_s}$ describes the z-orientation of the sensors on the vehicle.

As soon as several ($N_s \geq 2$) radar sensors are used on the same mobile platform, the equations can be transformed into a common system equation, and thus the velocity vector $\mathbf{v}_c$ can also be unambiguously determined:

$$\underbrace{\begin{bmatrix} \mathbf{v}_{r_1} \\ \mathbf{v}_{r_2} \\ \vdots \\ \mathbf{v}_{r_{N_s}} \end{bmatrix}}_{\mathbf{v}_r} = \underbrace{\begin{bmatrix} \psi_1 \\ \psi_2 \\ \vdots \\ \psi_{N_s} \end{bmatrix}}_{\psi} \cdot \underbrace{\begin{bmatrix} v_\omega \\ v_x \\ v_y \end{bmatrix}}_{\mathbf{v}_c}. \tag{11.14}$$

Equation (11.14) can then be solved for the sought velocity vector $\mathbf{v}_c$ using a Moore-Penrose inverse of ψ:

$$\mathbf{v}_c = \psi^+ \cdot \mathbf{v}_r. \tag{11.15}$$

The velocity of the vehicle can be determined for all three degrees of freedom. However, this requires stationary targets from which the velocity estimate is derived. Since this condition is rarely met in reality, non-stationary targets must be filtered out using a random sample consensus (RANSAC) algorithm. An example of such an evaluation is shown in Fig. 11.7 for a linear motion of the vehicle at a speed of 10 m/s. The detected targets are shown as *black circles*, while the targets filtered out by the RANSAC algorithm as stationary targets have an additional *red fill*. Based on all stationary targets within a threshold (*turquoise area*), the ego-motion estimation is performed according to (11.14). The integration of this velocity information over time results in a position estimate relative to the starting point for each measurement.

Dead reckoning is prone to cumulative errors. Any error in the estimate of position is accumulated over time, leading to increasing uncertainty about the actual position.

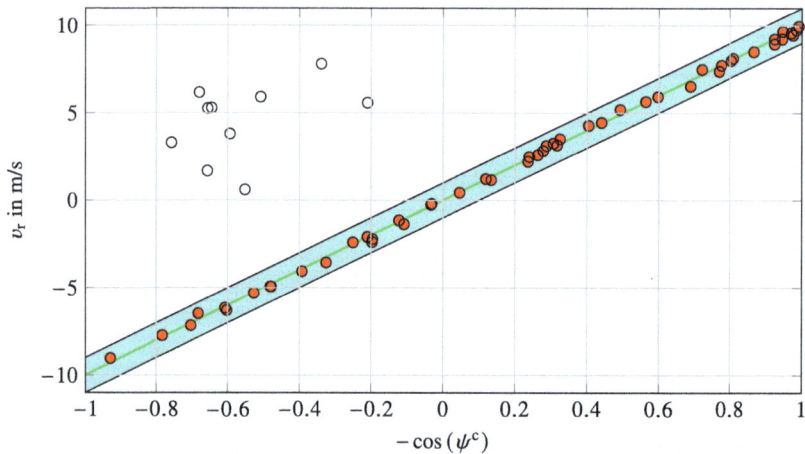

Fig. 11.7 Illustration of ego-motion estimation based on target lists of distributed radar sensors for a vehicle speed of 10 m/s. The detected targets are shown as *black circles*, whereas those detected targets that correspond to the most likely velocity model are *filled in red*. The threshold for identifying the stationary targets using the RANSAC algorithm is shown in *turquoise*

11.4 Simultaneous Localization and Mapping (SLAM)

Small inaccuracies in the velocity estimate or in the starting position can add up significantly over time, resulting in a so-called trajectory drift. Therefore, this type of self-localization is typically only used for short trajectories, whereas for longer trajectories, the position estimation is based on scan matching methods.

11.4.2 Scan Matching

Scan matching refers to the procedure by which the vehicle's current sensor information is compared with previous measurements to estimate the current position. This is achieved by seeking the best match between the current sensor scan and previous scans to determine position and orientation. This is schematically shown in Fig. 11.8. Once the same scene, in Fig. 11.8 for example a wall, is captured by the radar sensor in two consecutive time steps $t \rightarrow t+1$, the rotation and translation of the vehicle can be determined from this.

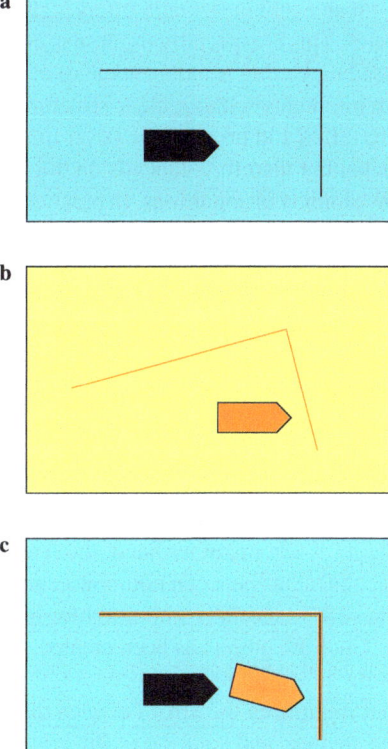

Fig. 11.8 Illustration of the scan-matching principle for two different measurements and two different vehicle poses; **a** Environment in the vehicle coordinate system at the time t. **b** Environment in the vehicle coordinate system at the time $t+1$. Due to the representation in the vehicle coordinate system, the change in pose of the vehicle corresponds to an inverse translation and rotation of the environment. This changed environment is measured by the radar sensors. **c** Estimated vehicle position at time $t+1$ in relation to the vehicle position at time t. With the help of scan matching, the change in pose of the vehicle is determined based on the changed relative environment

The classic structure of a scan matching algorithm consists of several steps that are used to estimate the current position and orientation of a robot or vehicle based on sensor information:

Feature extraction Features are extracted from the sensor data. Features are characteristic points or structures in the scene that are relevant for determining the position. These can be, for example, edges of objects or prominent points. The OGMs described in Sect. 11.2 provide a robust basis for this.

Creating a reference database A reference database contains representative features of the environment and their positions. This database serves as a reference for comparison with the current sensor data.

Pose estimation The actual scan matching process takes place in this step. The features extracted from the current sensor data are compared with those in the reference database. The aim is to find the best match (translation and rotation) for the extracted features. The most probable match corresponds to the most probable pose.

Various algorithms and approaches for scan matching are known from the literature, including iterative closest point (ICP), iterative closest line (ICL), as well as various probabilistic methods. Depending on the application and the available sensor data, the respective algorithms have different advantages and disadvantages.

Scan matching usually offers higher accuracy than dead reckoning for long trajectories. This is especially the case in environments with easily recognizable structures. Dead reckoning is independent of external environmental features and also works in environments without clear structures. However, it is more susceptible to cumulative errors and uncertainties over time. The choice between scan matching and dead reckoning therefore depends on the specific requirements of the application and the environmental conditions. In practice, hybrid approaches are often used that combine several self-localization methods to exploit the advantages of different techniques and compensate for their disadvantages. A graph-based SLAM fuses the information from dead reckoning and scan matching to ensure robust localization information for both short and long trajectories. The nodes of such a graph represent the vehicle poses at different measurement times, whereas the edges between the nodes represent the dependencies of the respective poses. The nodes are shown as *pentagons* in Fig. 11.9 and the dependencies as *arrows*.

The dependencies can be obtained in different ways. On the one hand, dependencies between two consecutive nodes exist based on dead reckoning and on the basis of scan matching. These kinds of dependencies are shown as *blue arrows* in Fig. 11.9. On the other hand, the features of scan matching can be used to establish dependencies between non-consecutive nodes, shown as *orange arrows* in Fig. 11.9. These are especially important for minimizing trajectory drifts for long trajectories.

Once the graph has been created, which is referred to as the front end, it is solved using an optimization procedure, which is the back end. This results in a solution that minimizes the error between all dependencies. A more detailed description of the creation and solution of the graph and of a possible scan matching algorithm is provided in [6].

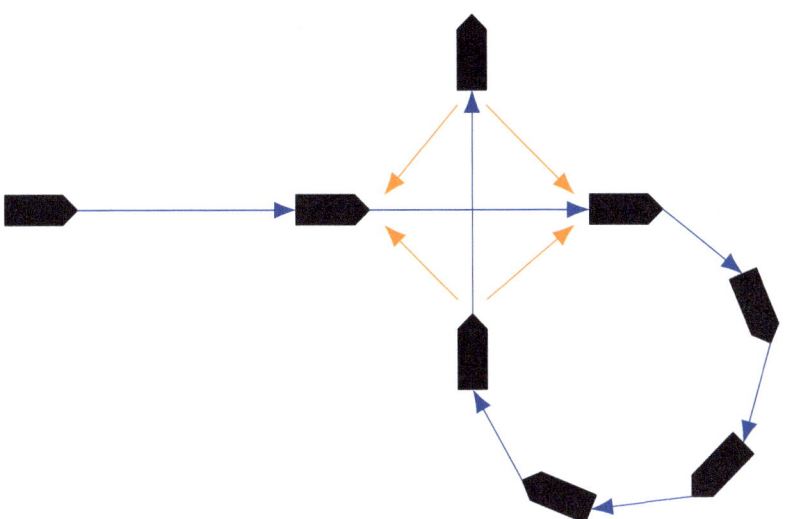

Fig. 11.9 Sketch of a graph with eight nodes representing the vehicle poses (*black*). The constraints are shown as arrows, with the *orange* arrows representing the cross-frame dependencies

References

1. X. Li, Z. Sun, Z. He, Q. Zhu, D. Liu, A practical trajectory planning framework for autonomous ground vehicles driving in urban environments, in *IEEE Intelligent Vehicles Symposium (IV)*, (2015), pp. 1160–1166. https://doi.org/10.1109/IVS.2015.7225840
2. H. Moravec, A. Elfes, High resolution maps from wide angle sonar, vol. 2 (1985), pp. 116–121. https://doi.org/10.1109/ROBOT.1985.1087316
3. P. Hügler, T. Grebner, C. Knill, C. Waldschmidt, UAV-borne 2-D and 3-D radar-based grid mapping. IEEE Geosci. Remote. Sens. Lett. 1–5 (2020). https://doi.org/10.1109/LGRS.2020.3025109
4. J. Degerman, T. Pernstål, K. Alenljung, 3D occupancy grid mapping using statistical radar models, in *IEEE Intelligent Vehicles Symposium (IV)*, (2016), pp. 902–908. https://doi.org/10.1109/IVS.2016.7535495
5. S. Thrun, Learning occupancy grids with forward models, in *International Conference on Intelligent Robots and Systems*, vol. 3 (2001), pp. 1676–1681. https://doi.org/10.1109/IROS.2001.977219
6. T. Grebner, R. Riekenbrauck, C. Waldschmidt, Simultaneous localization and mapping (SLAM) for synthetic aperture radar (SAR) processing in the field of autonomous driving. IEEE Trans. Radar Syst. 1–1 (2023). https://doi.org/10.1109/TRS.2023.3347734

Open Access This chapter is licensed under the terms of the Creative Commons Attribution 4.0 International License (http://creativecommons.org/licenses/by/4.0/), which permits use, sharing, adaptation, distribution and reproduction in any medium or format, as long as you give appropriate credit to the original author(s) and the source, provide a link to the Creative Commons license and indicate if changes were made.

The images or other third party material in this chapter are included in the chapter's Creative Commons license, unless indicated otherwise in a credit line to the material. If material is not included in the chapter's Creative Commons license and your intended use is not permitted by statutory regulation or exceeds the permitted use, you will need to obtain permission directly from the copyright holder.

Chapter 12
Synthetic Aperture Radar (SAR) for Millimeter Wave Applications

If an application requires angular resolution capabilities, radar images or environment models that have a significantly better resolution than what is possible with MIMO radars or grid mapping approaches, the principle of synthetic aperture radar (SAR) can be utilized. In many applications in robotics and autonomous driving, very accurate environment models or radar images with resolutions in the range of a few centimeters are required. Radar images with millimeter-range resolutions are also used in automation and safety engineering applications. In all these applications, the SAR principle is used.

In the millimeter wave range, resolution in the range dimension can be improved by using larger bandwidths, as described in Sect. 3.3. However, to improve resolution orthogonal to the range direction, large apertures are necessary. This is where the SAR principle comes into play when high-resolution angle estimation methods and MIMO radars are no longer sufficient.

SAR enables the generation of a synthetically enlarged aperture with high angular resolution by sequential spatial sampling the beat signal. This is done by utilizing the motion of the transmitter and receiver of a radar platform that, for example, is attached to a moving vehicle or robot and moves along a known trajectory. Alternatively, radar sensors are mounted on linear rails, for example in automation technology, so that the movement of the radar sensor allows a spatial synthesis of the aperture. Thus, SAR improves the quality of the angle estimation without using additional hardware channels.

In general, the SAR principle can be reversed so that the radar sensor is stationary and the target moves on a known trajectory, e.g. on a conveyor belt. This reversed principle is referred to inverse SAR, but this topic is not discussed further in this book.

In SAR processing, the sequentially sampled beat signal is synthesized along the aperture, which generates the SAR image. This image represents the information that would have been obtained from a real aperture of the length of the sampling path.

The *synthetic aperture* therefore refers to the scanning path over which the signal is synthesized or, respectively, sampled.

A fundamental difference between SAR and grid maps from Chap. 11 lies in the information used to generate the images and environmental models. For grid maps, target lists are used, i.e., a conventional signal processing chain with CFAR and angle estimation (see Chaps. 5 and 7) has already been carried out. In contrast, SAR is based on raw data, i.e., a conventional radar signal processing chain is not required.

12.1 Fundamentals and Resolution

A SAR scenario is shown as an introductory example in Fig. 12.1 for a vehicle with a side-looking radar sensor. For reasons of clarity, this radar sensor is equipped with only one antenna. At specific, often regular points in time t, the radar sensor transmits a signal that is reflected at surrounding targets and then received. Due to the motion of the vehicle, the radar sensor is at a different position at each measurement point, which leads to spatial sampling or, respectively, to a synthetic aperture. In the simple case of a linear trajectory with constant vehicle motion, the synthetic aperture is similar to that of a large ULA. In SAR imaging, the dimension in the direction of motion is referred to as the azimuth direction and the dimension orthogonal to it as the range dimension. The resolution in range direction is determined by the

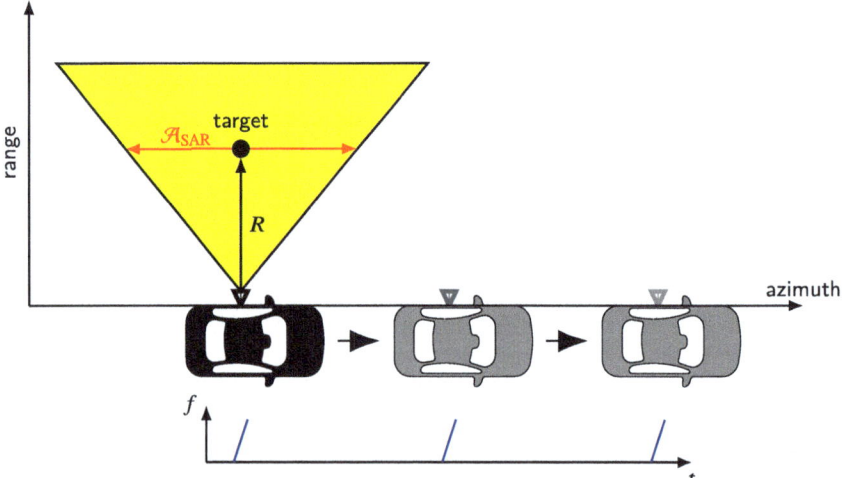

Fig. 12.1 Visualization of three measurement positions of the synthetic aperture; The vehicle moves along a known trajectory and emits a chirp at each of the three measurement intervals. By driving the vehicle to different positions, a synthetic aperture is created that allows the target to be captured from different angles. The coherent processing of this information is referred to as SAR. The FoV of the radar is depicted in *yellow*

12.1 Fundamentals and Resolution

bandwidth B of the radar signal and the viewing angle to the imaging strip. The resolution in azimuth direction is determined by the length of the synthetic aperture, analogous to real apertures. Assuming a real physical aperture, the angle resolution is approximately

$$\Delta\psi \approx \frac{\lambda}{\mathcal{A}}, \tag{12.1}$$

where $\mathcal{A}$ corresponds to the aperture length. This approximation is obtained by linearizing the cosine function in (6.25) in Sect. 6.6.1 and using the aperture $\mathcal{A} = N_{\text{ant}}\Delta\psi$. In SAR imaging, the synthetic aperture replaces the physical or real aperture. This circumstance is mathematically described by

$$\Delta\psi_{\text{SAR}} \approx \frac{\lambda}{2\mathcal{A}_{\text{SAR}}}. \tag{12.2}$$

While in the case of a real aperture, the transmitter is assumed to be stationary at a fixed location and the phase difference of the beat signals results exclusively from the spacing between the receiving antennas, in the case of SAR, the transmit and receive antennas move simultaneously. This leads to a doubling of the phase difference, from which the factor 2 in (12.2) follows. Thus, the azimuth or angular resolution in SAR imaging results from the size of the synthetic aperture $\mathcal{A}_{\text{SAR}}$. This aperture is determined by the distance that the radar can travel without the target object leaving its FoV. In Fig. 12.1, this area is highlighted in *yellow* and the maximum aperture length for an exemplary target is depicted in *red*.

The resolution in the image plane Δ_{SAR} follows from (12.2) by means of a small-angle approximation (linearization of the cosine function):

$$\Delta_{\text{SAR}} = \frac{\mathcal{A}}{2}. \tag{12.3}$$

Thus, the resolution in the image plane for SAR is independent of the range. Furthermore, it can be seen that the resolution at the target position, or in the image plane, depends only on the real aperture size. The larger the antenna, the more focused the beam of the antenna and the shorter the path length within which the target is in the FoV of the antenna while the sensor is moving, thus limiting the synthetic aperture. In contrast to the angle estimation algorithms from Chap. 7, a small antenna with a large opening angle is advantageous for SAR processing, at least if a sufficiently good SNR is available despite small antenna gains.

To achieve these high resolutions in practice, many different boundary conditions must be met. Among other things, the synthetic aperture must be sampled very accurately according to the sampling theorem. This is very challenging in many applications, especially in the millimeter wave range, since the spatial error in the sampling must be only fractions of the wavelength. If the trajectories are not sufficiently well known, as is often the case with robots or vehicles, large spatial sampling errors arise, leading to blurred SAR images with a lot of clutter. However, a radar-based motion estimation, as presented in Sect. 11.4, makes it possible to almost achieve the theoretical resolution limits [1].

12.2 SAR Processing

SAR processing refers to the generation of radar images by coherently processing radar data acquired. In the following, the use of a chirp sequence radar is assumed as an example, since this is the predominant radar type in the millimeter wave range. SAR processing for other modulation methods is the same, provided that they allow range measurements.

Basically, the SAR processing is a kind of angle estimation for which, in many applications, neither the far-field condition (due to large apertures) nor equidistant antenna spacings (due to non-constant motion) are guaranteed. There are two main differences compared to conventional angle estimation: Firstly, the angle estimation is not carried out on the basis of an entire radar measurement, i.e. a chirp sequence frame, but based on each individual ramp or chirp that could theoretically detect the target. Secondly, the angle estimation is not carried out for each target detected by a CFAR, but for all range cells. For this reason, the term imaging is often used instead of angle estimation.

Due to the high integration gain of SAR processing, it is possible to detect weakly reflecting targets whose SNR would be too low when processing a single chirp or measurement frame. However, the high integration gains can only be achieved if all measurement data can be acquired coherently via a synthetic aperture. This results in the high demands on spatial sampling. If the sample locations are not sufficiently known, the individual measurements or, respectively, samples cannot be added coherently, which significantly reduces the integration gain.

There are many different methods for generating SAR images. Generally, these are divided into algorithms in the frequency domain and algorithms in the time domain. The advantage of the time-domain algorithms lies in their universal field of application, since many of these approaches do not require the trajectory to be linearized. Regardless of the algorithm used, the basic principle is often identical. For each cell ϱ in the environment $\mathbf{U}$ to be mapped, the measured range information is phase-corrected according to the range to be evaluated (to cell ϱ). This corresponds to a matched filter-based phase correction. If the cell ϱ to be evaluated represents a target, the signals of all L ramps add up constructively after phase correction. If the cell represents noise, the phases of the corrected signals are random and do not add up constructively. The phase correction, which is similar to the application of the steering vector in angle estimation, is calculated individually for each pixel and sensor position without approximations, enabling the generation of SAR images without artifacts, even for curved trajectories.

For SAR applications in the millimeter wave range, the so-called backprojection (BP) algorithm is often used. It is a time-domain algorithm and does not require

12.2 SAR Processing

linear trajectories. The starting point for this is the sampled beat signal **y** of a chirp or an FMCW ramp, which is transformed into the frequency domain by taking the DFT with

$$\mathbf{Y} = \mathcal{F}\{\mathbf{y}\}. \tag{12.4}$$

The frequency vector **Y** assigns a complex-valued amplitude to each discrete range cell of the range vector **R**. In the case of a complex-valued sampling and subsequent DFT without *zero padding*, the corresponding range vector is given by

$$\begin{bmatrix} \mathbf{Y} \\ \mathbf{R} \end{bmatrix} = \begin{bmatrix} Y_1 & Y_2 & \ldots & Y_K \\ 0 & \Delta R & \ldots & \Delta R \cdot (K-1) \end{bmatrix}^\mathrm{T}. \tag{12.5}$$

ΔR corresponds to the range resolution of the radar sensor, according to (5.32). Based on the range information **R** and the associated complex amplitudes **Y**, a SAR processing using the BP algorithm is performed for the area to be mapped **U**. In this process, each of the cells is attributed a received amplitude $A(\varrho)$. This complex value is calculated iteratively based on all measurements ($\mathbf{\Lambda}$) and all positioning data ($\mathbf{P}_\mathrm{F}^\mathrm{g}$) using the BP algorithm:

$$A\left(\varrho^\mathrm{g} \,\middle|\, \mathbf{\Lambda}_{1:t}, \mathbf{P}_{\mathrm{F}_{1:t}}^\mathrm{g}\right)$$

$$= A\left(\varrho^\mathrm{g} \,\middle|\, \mathbf{\Lambda}_{1:t}1, \mathbf{P}_{\mathrm{F}_{1:t}}^\mathrm{g}1\right) + A\left(\varrho^\mathrm{g} \,\middle|\, \mathbf{\Lambda}_t, \mathbf{P}_{\mathrm{F}_t}^\mathrm{g}\right) \tag{12.6}$$

$$= A\left(\varrho^\mathrm{g} \,\middle|\, \mathbf{\Lambda}_{1:t}1, \mathbf{P}_{\mathrm{F}_{1:t}}^\mathrm{g}1\right) + Y\left(\varrho^\mathrm{g} \,\middle|\, \mathbf{\Lambda}_t, \mathbf{P}_{\mathrm{F}_t}^\mathrm{g}\right) \cdot \exp\left(\mathrm{j} \cdot \phi_{\mathrm{BP}}\left(\varrho \,\middle|\, \mathbf{P}_{\mathrm{F}_t}^\mathrm{g}\right)\right). \tag{12.7}$$

The received amplitude of the cell $A(\varrho^\mathrm{g} \mid \mathbf{\Lambda}_{1:t}, \mathbf{P}_{\mathrm{F}_{1:t}}^\mathrm{g})$ up to the measurement time t is composed of the received amplitude of the cell $A(\varrho^\mathrm{g} \mid \mathbf{\Lambda}_{1:t}1, \mathbf{P}_{\mathrm{F}_{1:t}}^\mathrm{g}1)$ from all previous times $1{:}t-1$ and the phase-corrected value of the current measurement $Y(\varrho^\mathrm{g} \mid \mathbf{\Lambda}_t, \mathbf{P}_{\mathrm{F}_t}^\mathrm{g})$. The assignment between the cell to be evaluated ϱ and the corresponding range cell of the vector **Y** is achieved using a *nearest neighbor* search. The term

$$\phi_{\mathrm{BP}}\left(\varrho \,\middle|\, \mathbf{P}_{\mathrm{F}_t}^\mathrm{g}\right) = -\mathrm{j}2\pi \frac{f_c}{c} \cdot 2R\left(\varrho \,\middle|\, \mathbf{P}_{\mathrm{F}_t}^\mathrm{g}\right). \tag{12.8}$$

describes a range-dependent phase correction to ensure constructive superposition for targets and random superposition for noise. This ultimately corresponds to a conventional angle estimation without a far-field approximation.

This procedure is applied to each of the N_ϱ cells of the rasterized map **U** and for all L ramps of each frame. The movement of the vehicle results in a constantly changing sensor position, creating a large synthetic aperture.

The steps of the BP algorithm are illustrated in Fig. 12.2 for a radar sensor with one transmit and receive antenna; and for three range cells, depicted as three shades of gray. For each new measurement or ramp, each cell within the FoV is first assigned the corresponding complex amplitude, as shown in Fig. 12.2(1a). Since only three range cells are considered, only three different phasors are registered in the map. Next, an individual phase correction $\phi_{\mathrm{BP}}(\varrho \mid \mathbf{P}_{\mathrm{F}_i}^{\mathrm{g}})$ (12.8) is calculated for each cell. After applying this phase correction, the image shown in Fig. 12.2(1b) is obtained. The phases of each cell are now phase-corrected according to the range of the cell. The two cells highlighted in *red* illustrate that although the phase of the measurement is identical for these two cells, as shown in Fig. 12.2(1a), a different range between the cell and the radar sensor results in a different phase correction. Then the platform and with it the antenna is moved and the same principle is repeated. This is illustrated in Fig. 12.2(2), where the antenna and thus its FoV were moved one cell to the right. The targets are localized using the sum of information from these two cells, which results in the Fig. 12.2(3). It can be seen from the figure that the sum leads to constructive superposition for some cells and destructive superposition for other cells. These extreme cases are shown as *red squares*. If a cell has the same phase for all ramps to be evaluated after the phase correction, a constructive superposition occurs and thus a high amplitude (*yellow*) in the resulting SAR figure. If the opposite is the case, and thus a phase offset of 180° is given, a destructive superposition occurs. These cells are shown in *blue*. The more frames are evaluated, the clearer the distinction between occupied and non-occupied cells.

Figure 12.3a shows a SAR image generated from the same measurement data as the grid maps in Fig. 11.5 in Sect. 11.3. Compared to both the AGM and the OGM, the significantly increased resolution of the SAR image is evident. The vehicles can be clearly identified as such. The high resolution is also evident in the trees and, in particular, in the branches, which can be clearly distinguished from each other. In addition to the high resolution, the high integration gain is a further advantage of SAR. This allows weak-reflecting targets, such as the wooden fence at $y=-8$ m, to be clearly imaged, which is not possible in this quality and robustness with AGMs or OGMs. While the OGM only visualizes the posts and the AGM is too blurry to separate the fence from the surrounding area, the fence is clearly identifiable in the SAR image. Furthermore, SAR offers the possibility of distinguishing the roadway from gravel parking lots due to the high integration gain. The road junction at $x=76$ m illustrates this. However, a disadvantage of conventional SAR processing is the strong amplitude dependence that also occurs in AGMs.

12.3 Probabilistic SAR Processing

Amplitude-based SAR processing, as described in Sect. 12.2, has a significant disadvantage despite the large aperture and the resulting high-resolution images – the amplitude dependency. Similar to AGMs, the targets are evaluated based on their amplitude. This leads to the vehicles at the upper edge of the image in Fig. 12.3a

12.3 Probabilistic SAR Processing

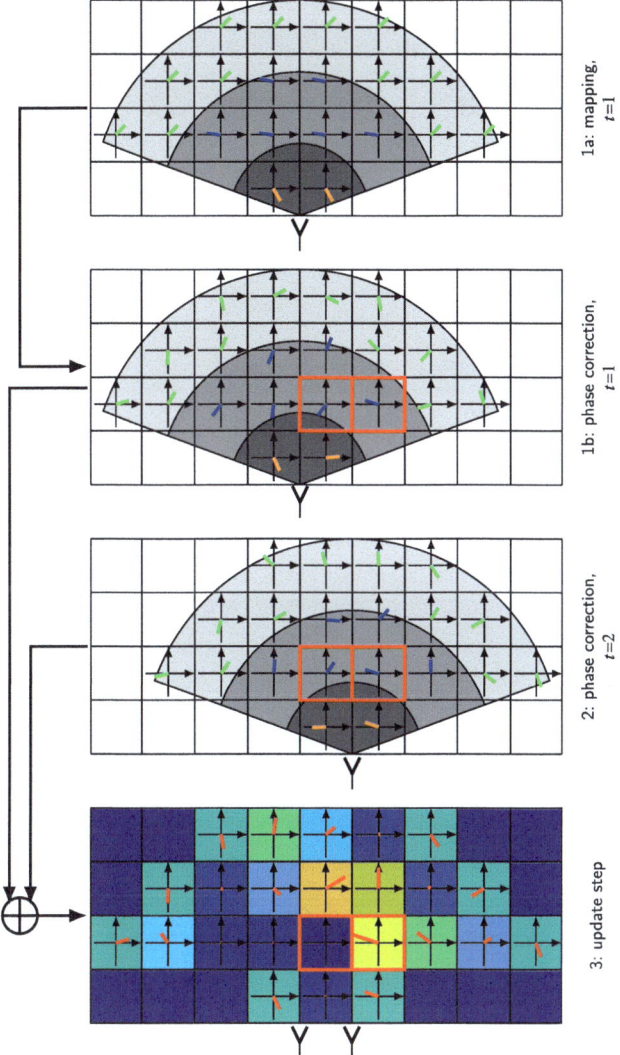

Fig. 12.2 Visualization of the BP algorithm for two synthetic antenna positions to be evaluated. At time $t=1$, a measurement is performed. The measurement data is processed and the phase and amplitude information of the three exemplarily chosen range cells are entered into the corresponding cells of the map. Subsequently, a phase correction is applied based on the distance between the sensor and the cell in the map. At time $t=2$, a new measurement is performed, the data is processed and a phase correction is performed. Finally, the map from the first measurement and the map from the second measurement are fused by summation. This coherent processing results in a coherent superposition of the complex-valued pointers for targets and a random superposition of the complex-valued pointers for cells that do not represent a target

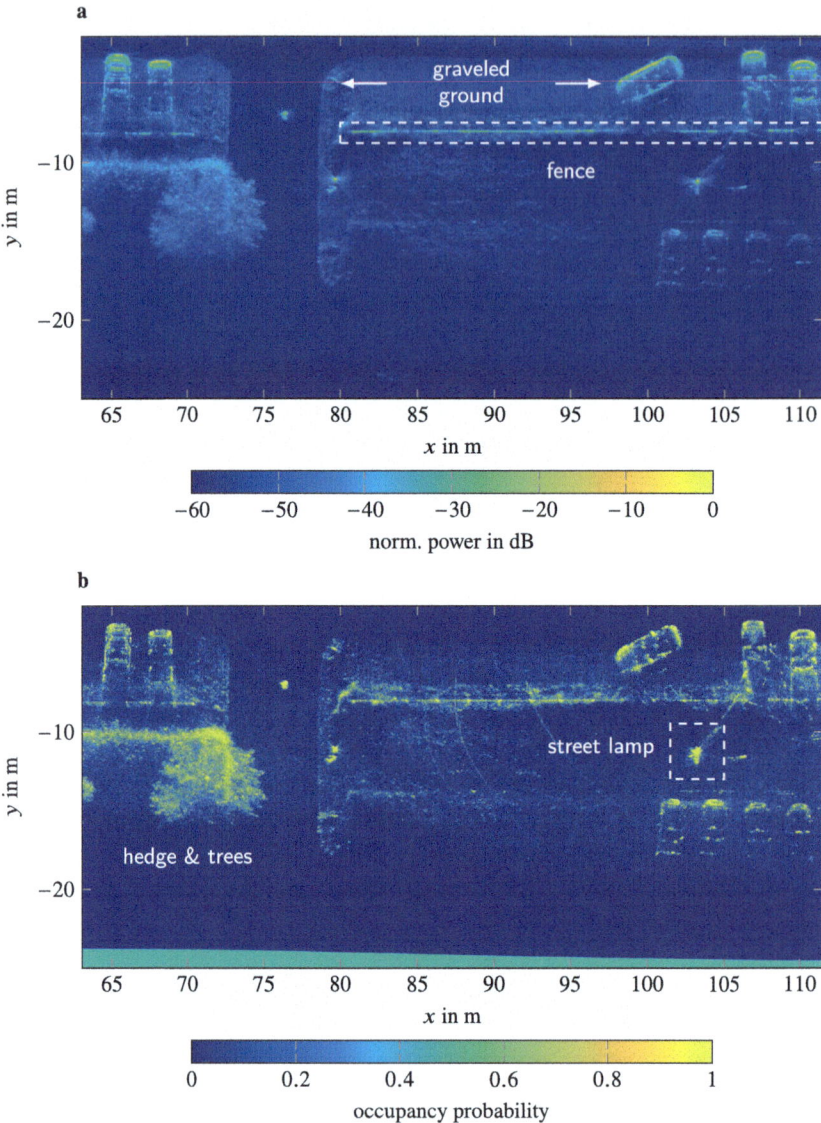

Fig. 12.3 Visualization of **a** an amplitude-based SAR mapping and **b** a probabilistic SAR mapping. Relevant targets such as trees, hedges, street lamps, as well as gravel and wooden fences are highlighted

($y = -2$ m) appearing almost 50 dB stronger than the vehicles in the bottom half of the SAR-image ($y = -18$ m). However, the actual purpose of a map of the surrounding area is to assess the probability of individual cells being occupied, not to display the received reflection amplitude of individual targets.

12.3 Probabilistic SAR Processing

In contrast, probabilistic SAR processing ensures an almost amplitude-independent image of the environment, as explained in [2]. The basic structure of the BP algorithm similar to (12.7) is retained, since a range-proportional phase correction of the measured values must still be carried out. The key difference is that complex probabilities are processed instead of complex amplitudes. This requires probabilistic target and phase models, which are then fused using a probabilistic update function. A detailed description of the algorithm can be found in [2].

The result of such a probabilistic SAR processing is shown in Fig. 12.3b. In comparison to the amplitude-based SAR processing in Fig. 12.3a, the entries in the map are independent of the received signal amplitude and thus independent of the range. Instead, each cell is assigned an occupancy probability. As a result, vehicles at $y=-2$ m have almost the same probability of occupancy (100%) as vehicles at $y=-18$ m; the same applies to the wooden fence and the trees. The result is an extremely robust image of the environment, without sacrificing resolution. Both the contour of the vehicles as well as the separation between the road and the gravel

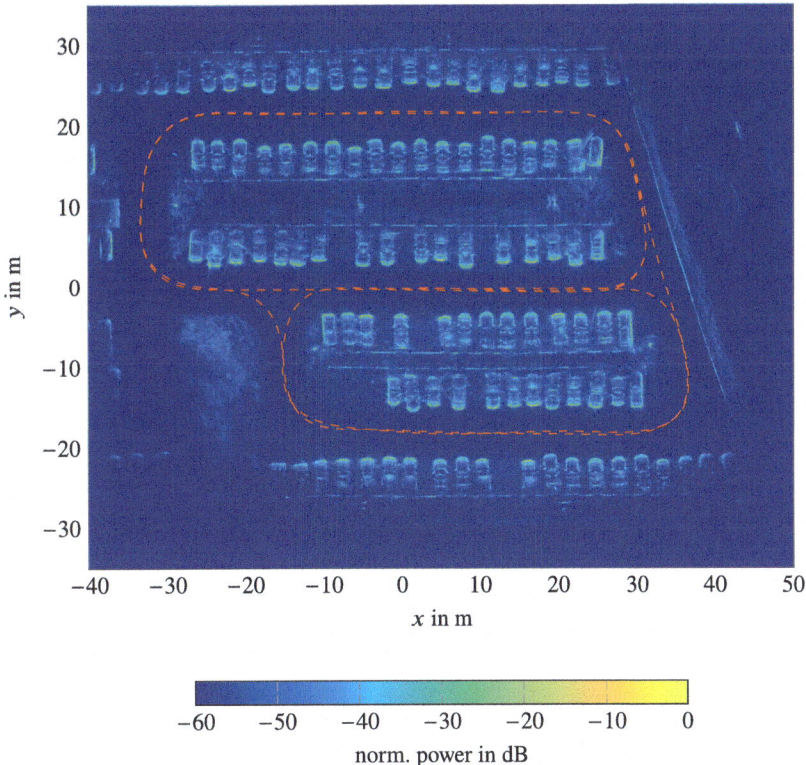

Fig. 12.4 SLAM-based mapping of a parking lot using SAR-based processing at 77 GHz. The *red dashed* line shows the trajectory of the vehicle

parking lot are retained. Probabilistic SAR-processing thus combines the advantages of SAR, in particular the high resolution, with the advantages of OGMs, such as the probabilistic calculation and representation as well as the resulting amplitude independence.

12.4 SLAM for SAR Applications

The SLAM approach described in Sect. 11.4 is not only applicable to create grid maps, but also to create high-resolution SAR images. In this case, localization is based solely on radar data, eliminating the need for external localization using GNSSs or inertial measurement units (IMUs). An autonomous high-resolution mapping of a parking lot is thus possible, as shown in Fig. 12.4. For the SAR processing in the figure, the two sensors 2 and 4 (side-looking) were used, as shown in Fig. 11.6.

Compared to Fig. 12.3a, an identical image quality can be achieved. The fences and the vehicles are still clearly visible. Furthermore, the gravel can be clearly distinguished from the road. Although the localization was derived exclusively from the radar data, no SNR loss or, respectively, image dynamic loss is visible compared to a localization using GNSS or other external sensors.

References

1. Y.E. Ritterbusch, J. Fink, C. Waldschmidt, Indoor synthetic aperture radar measurements of point-like targets using a wheeled mobile robot, in *European Conference on Synthetic Aperture Radar (EUSAR)*, (2024), pp. 595–600
2. T. Grebner, A. Grathwohl, P. Schoeder, V. Janoudi, C. Waldschmidt, Probabilistic SAR processing for high-resolution mapping using millimeter-wave radar sensors. IEEE Trans. Aerosp. Electron. Syst. **59**(5), 4800–4814 (2023). https://doi.org/10.1109/TAES.2023.3289784

Open Access This chapter is licensed under the terms of the Creative Commons Attribution 4.0 International License (http://creativecommons.org/licenses/by/4.0/), which permits use, sharing, adaptation, distribution and reproduction in any medium or format, as long as you give appropriate credit to the original author(s) and the source, provide a link to the Creative Commons license and indicate if changes were made.

The images or other third party material in this chapter are included in the chapter's Creative Commons license, unless indicated otherwise in a credit line to the material. If material is not included in the chapter's Creative Commons license and your intended use is not permitted by statutory regulation or exceeds the permitted use, you will need to obtain permission directly from the copyright holder.

Chapter 13
Coexistence and Interference of Radar Sensors

Due to the widespread use of radar sensors in the millimeter wave range, multiple sensors are often used in close proximity or in the same application scenario. This can lead to interference between the sensors. In road traffic in particular, many vehicles are nowadays equipped with one or more radar sensors, so that in dense traffic scenarios, interference can occur between sensors on different vehicles.

This chapter first presents an overview on interfered signals in radar applications in Sect. 13.1 and then provides a more specific description of the impact of interference on analog and digital radars in Sect. 13.2. A selection of countermeasures is presented in Sect. 13.3.

13.1 Interfering Signals

Generally, interference between radar sensors can only occur if several sensors operate at the same time and in the same frequency band. Since all sensors in an application operate in the same frequency band due to frequency regulation, there is always an inherent risk of interference. In contrast to a regular radar measurement, in which the received signal is attenuated according to the radar equation with $1/r^4$, an interference signal may experience significantly less attenuation. Interference signals are often received directly on the LOS, rather than being reflected from a target in the channel. This results in an attenuation of only $1/r^2$ between the transmitter and the interfered receiver, i.e. interference signals can be significantly stronger than radar signals reflected from targets. In road traffic, for example, the interference level of a vehicle's radar sensor is often significantly higher than the signal level of the reflection from the other vehicle that is equipped with the interfering radar sensor. The probability that a radar experiences interference can be reduced if only narrow bandwidths, i.e. small sections of the available frequency band, are used and these frequency ranges are only occupied for short periods. However, the probability of occurrence says nothing about the severity of the effects of interference. Short interference with a high level of interfering signal power from analog radars can have the same effect as permanent interference from digital radars with a low level of interfering signal power.

13.2 Impact of Interference

If a radar sensor is interfered by the transmit signal of another radar sensor, the interfering signal can be received and mixed into the baseband. This usually results in an increase of the noise floor, which is examined in detail below. In principle, interference can cause the receiver to be saturated. However, this hardly occurs in practice due to the high signal level required for this. Although in theory it is possible for interference to generate false targets, in practice this hardly occurs either. This is because the interferer and the receiver of the interfered radar sensor are not coherently coupled, i.e. they operate with different oscillators. This gives rise to different drifts and noise in the oscillators, causing any target peaks in the baseband to be smeared, so that no sharp target peaks can arise.

13.2.1 Impact on Analog FMCW and Chirp Sequence Radars

If an FMCW radar sensor, i.e. a linear frequency modulated sensor, is interfered by another linear frequency modulated sensor, the frequency ramps of the interferer and the radar are mixed together. In the event that both the interferer and the radar have the same ramp slopes, theoretically a false target may be generated. However, the different behavior of the oscillators ensures that in practice no target is generated. It is much more likely that the two systems have different ramp slopes, as shown in Fig. 13.1. As a result, the frequency of the beat signal changes over the frequency ramp. Since only those frequency components of the beat signal that are not filtered out by the lowpass or bandpass filter before the ADC are present at the receiver output, the interference signal is present only for a very short time

$$T_{\text{int}} = \frac{2 B_{\text{beat}}}{\frac{B_{\text{jam}}}{T_{\text{jam}}} - \frac{B}{T}}, \tag{13.1}$$

where the subscript text "jam" stands for "jammer", equal to the interferer. The more similar the slopes of the frequency ramps are, the longer T_{int} is. Although T_{int} is usually only a fraction of the ramp duration, a high interference power level can be received during this time. The interfered beat signal stemming from the mixing has a phase response that follows a parabola

13.2 Impact of Interference

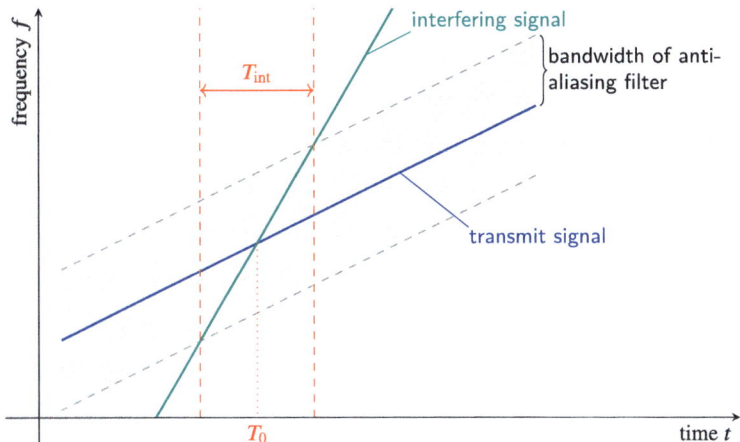

Fig. 13.1 The interfering signal (*green*) disturbs an FMCW radar sensor. In the sensor, the transmitted signal (*blue*) is mixed with the interfering signal. However, the anti-aliasing filter filters the output signal of the mixer. Signal components of the output signal that lie outside the filter bandwidth (*gray area*) are eliminated. Consequently, the interference occurs only very briefly for the duration T_{int} (*red*)

$$y(t) = A_{\text{int}} \cos\left(2\pi \left(f_{c\,\text{jam}} - f_c\right) t + \pi\left(\frac{B_{\text{jam}}}{T_{\text{jam}}} - \frac{B}{T}\right) t^2 + \phi_0\right) \text{rect}\left(\frac{t - T_0}{T_{\text{int}}}\right). \tag{13.2}$$

Due to the Fourier transform, the short duration T_{int} of the interference causes the interference power to be smeared over the entire range spectrum, i.e. the SNR of the measurement is impaired, as illustrated in Fig. 13.2.

If several successive ramps are affected, the effect of the interference on the velocity spectrum depends on how stable the two oscillators of the two radar sensors are in comparison to each other [1]. As a rule, the interference power accumulates within a few velocity cells, so that only the corresponding sections of the R-v matrix are disturbed.

If a linear frequency modulated radar is disturbed by a digital OFDM radar, many subcarriers of the OFDM signal intersect with the frequency ramps. Each intersection can be interpreted as an interference of a frequency ramp with a phase-modulated CW radar, i.e. for each intersection similar interferences occur as described for the interference of linear frequency-modulated radars. The individual interferences of the many intersections add up to the total interference. Due to the modulation of the OFDM subcarriers, the individual interferences are uncorrelated and the total interference signal is distributed over the entire R-v matrix of the radar. The resulting SIR (signal-to-interference ratio) is given by the integration of the interference signal over the measurement duration and bandwidth

$$\text{SIR} \propto BTL, \tag{13.3}$$

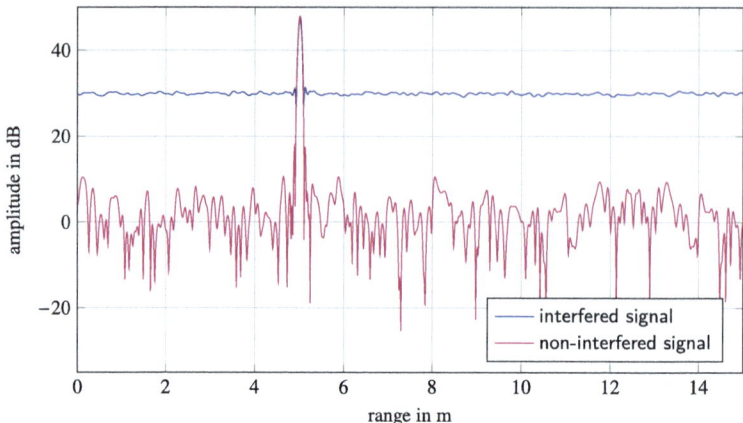

Fig. 13.2 Impact of interference in frequency domain on an FMCW or chirp sequence modulated radar for a scenario with a single target at a range of $R = 5$ m. The interference increases the noise level significantly across the whole range spectrum

where L is the number of frequency ramps accumulated in the integration and T the duration of one ramp.

When a PMCW-modulated radar disturbs a linear frequency modulated radar, homogeneously distributed interference power is received, integrated over the entire measurement duration and bandwidth of the radar sensor, and distributed equally in all range cells. Since the codes are used periodically over time with PMCW radars, there may theoretically be an integration of the interference energy in certain velocity cells if there is a deterministic interference over many ramps. However, due to the independent oscillators in the radar sensors, in practice there is usually a smearing of the interference signals over the entire R-v matrix. Equation (13.3) applies accordingly.

13.2.2 Impact on Digital Radars

Since, unlike linear frequency modulated analog radars, digital radars instantaneously occupy a large portion of the total bandwidth during the entire measurement, interference is much more likely to occur than with analog radars. However, digital radars have a significantly lower spectral power density than analog radars in the same application context. This means that the received interference power, which is given by integrating the power density over the bandwidth, is of the same order of magnitude.

However, modern types of digital radar, such as stepped OFDM, do not use the entire bandwidth instantaneously during the entire measurement period, but only sections of it. This consequently reduces the probability of interference being encountered.

13.3 Countermeasures

As with analog radars, interference usually leads to an increase in the noise power level in the R-v matrix, i.e. to a deterioration of the SNR, since the individual signal components of the interference are uncorrelated in the receiver. Only temporally periodic interferences, e.g. by using cyclic codes in PMCW or repeated-symbol-OFDM, can lead to an integration of the interference power in certain velocity cells.

If digital radars are interfered by analog, linear frequency modulated radars, the interfering ramps can be clearly recognized in the spectrogram of the received signal, see for example Fig. 13.3.

13.3 Countermeasures

If radars in the millimeter wave range are disturbed, various methods are available to mitigate the interference or to repair the disturbed signal. The prerequisite for this is usually that the interfered radar sensor is not saturated by the interference. If saturation occurs and the receiver behaves non-linearly, almost all methods fail.

Countermeasures in the time or frequency domain are highly dependent on the modulation of the interfering signal. They are often designed for interference from analog radars. Almost only beamforming approaches are independent of the modulation scheme.

Before countermeasures can be taken, the interference usually has to be detected. A sudden sharp increase in the signal power level often serves as an indicator for this.

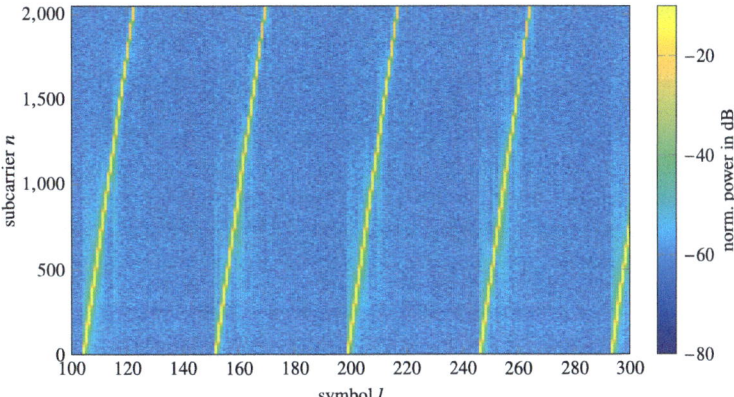

Fig. 13.3 Spectrogram of an OFDM signal that is disturbed by a chirp sequence radar signal within the same frequency band. The transmission power of the OFDM signal is distributed over a large bandwidth so that the power per subcarrier is low. In comparison, the entire transmission power of the chirp sequence signal is compressed to a small instantaneous bandwidth. Therefore, the chirp sequence ramps are clearly discernible in the OFDM spectrogram due to the higher instantaneous power level (*yellow stripes*)

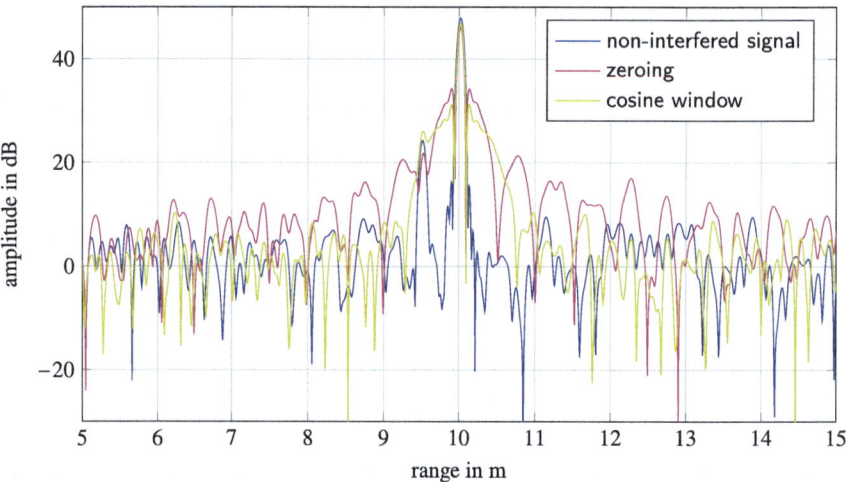

Fig. 13.4 Comparison of interference cancellation methods for a disturbed signal (simulated) with two targets at 9.5 m and 10 m. Zeroing (*magenta*) results in a significant suppression of the interfering signal components, but the side lobes of the targets are significantly increased. The cosine window (*yellow*) results in a reduction of the side lobes, but makes them wider

13.3.1 Measures in the Time Domain

Countermeasures in the time domain are applied to the digital time-domain signal before further steps in the signal processing chain are carried out.

As explained previously, the interference of two linear frequency modulated radars is of short duration; the interference is only present for the time T_{int}. A simple measure is therefore to set the affected signal section to zero. Mathematically, this is equivalent to multiplying the signal by a rectangular function, which results in convolving the signal in the frequency domain with a sinc function. Consequently, although the interference can be eliminated through this, the sinc function folds with the targets in the range spectrum, resulting in high side lobes that can mask weak targets.

Alternatively, instead of zeroing, the time domain signal can be smoothed at the disturbed samples, analogous to the application of a power-of-cosine window function of width T_F via

$$w(t) = \cos^2\left(\frac{\pi}{2} \cdot \frac{t}{T_F}\right). \qquad (13.4)$$

Smoothing reduces the side lobes of the sinc function, but broadens the main lobes. Figure 13.4 shows a comparison between an interference-free signal and a disturbed signal whose interfered samples has been set to zero or smoothed.

These fairly simple measures are effective in many cases. However, if interference persists for a long time, too much signal is eliminated. In this case, reconstruction approaches can be used to reconstruct the original signal. For example, the interference signal is estimated and subtracted from the disturbed signal [2]. The estimation of the interference signal is based on the prior knowledge that the interference has a parabolic phase response according to (13.2). Alternatively, compressed sensing methods are used. In this case, the disturbed signal section is set to zero and the resulting gap is reconstructed from the undisturbed signal components [3].

13.3.2 Measures in the Time-Frequency Domain

When a digital radar is disturbed by chirp sequence radars, the ramps are recorded in the frequency-time domain, as shown in Fig. 13.3 exemplarily. The interference can be eliminated by simply zeroing or smoothing the disturbed signal components. Figure 13.5 shows an example of this, in which an OFDM radar was interfered by a chirp sequence radar and the interference was then compensated for. After applying the countermeasure, the original SNR got almost restored.

13.3.3 Measures in the Frequency Domain

Countermeasures in the frequency domain aim to avoid interference by adaptively adjusting the utilized frequency spectrum. The goal is to avoid a temporal overlap of the spectrum used by the interferer and the victim radar. However, since in many applications radar sensors use a large fraction of the regulated spectrum, the possibilities for avoiding interference in the frequency domain are limited.

Random selection of the carrier frequency of a radar in the usable frequency range, as with frequency hopping, does reduce the probability of permanent interference. However, it does not reduce the probability of interference occurring.

It is more effective to change the carrier frequency only if interference is actually detected; and if there is no interference, the carrier frequency is not changed. In [4] it was shown for chirp sequence radar sensors that a change in carrier frequency is particularly effective if it does not happen randomly, but rather depending on the interference. In this case, a decrease in the carrier frequency is chosen if the ramps are disturbed in the upper frequency range. An increase in the carrier frequency is chosen if the ramps are disturbed in the lower frequency range.

In many cases, a change in carrier frequency does not completely avoid interference. However, the position of the intersections of the frequency ramps can be shifted so that the intersections are more likely to occur at the edges of the ramps, where the window functions of the classical signal processing strongly suppress the signal and thus the interference.

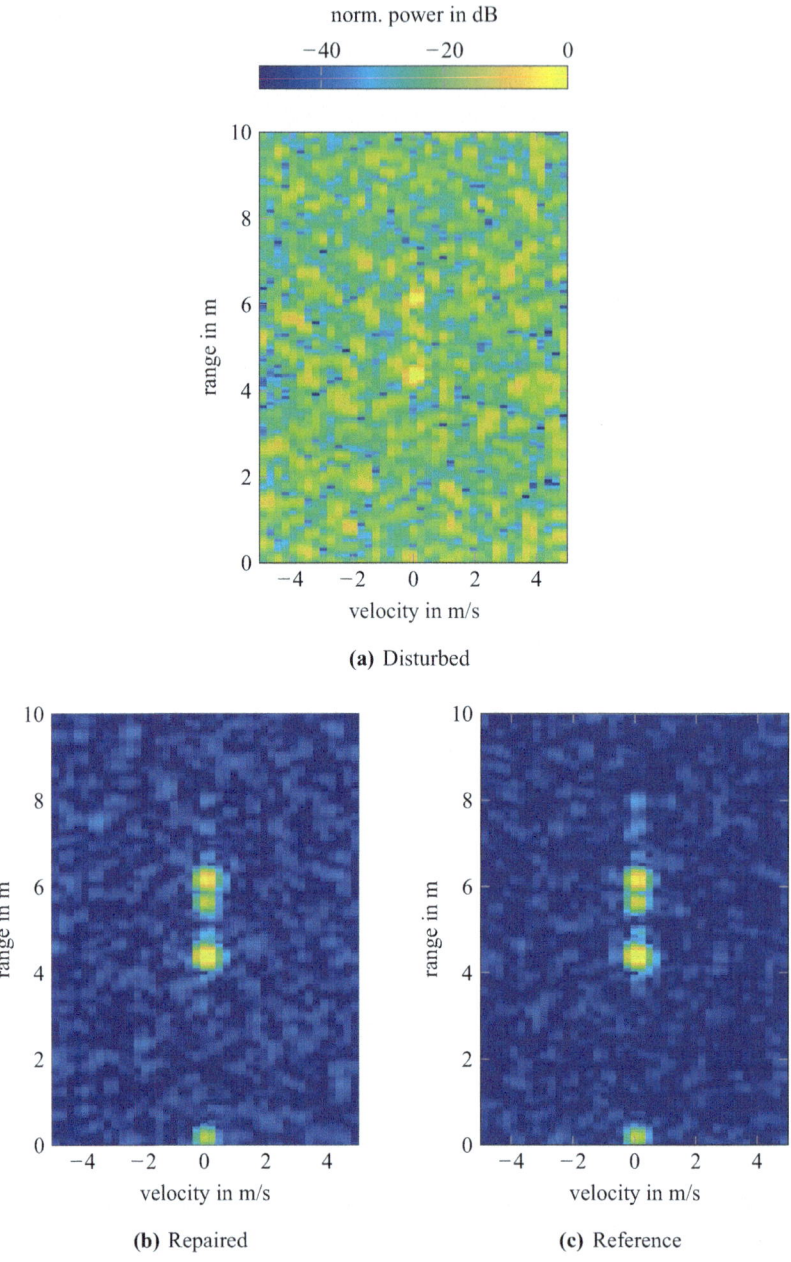

Fig. 13.5 Impact of interference mitigation on an OFDM radar measurement of three targets (4.25 m, 5.6 m, and 6.1 m) that is disturbed by a chirp sequence interferer. **a** R-v matrix obtained from the disturbed signal without interference mitigation. The targets are hardly discernible from the increased noise level. **b** R-v matrix after interference mitigation. The targets are clearly visible, the noise level is reduced and similar to that in **c** the non-disturbed reference measurement

13.3.4 Measures in the Angular Domain

Measures in the angular domain are based on beamforming techniques that steer a null in the receive pattern of the antenna array in the direction of the interferer. This does not automatically mean that the radar can no longer detect targets in this direction. If, for example, the interference power accumulates in one or a few velocity cells, it is sufficient to perform beamforming in these cells. In other velocity sectors, targets in the same angular direction as the interferer are still detectable.

Beamforming approaches for interference mitigation can be categorized into two groups: On the one hand, adaptive methods are used in which the angle of the interferer does not have to be explicitly known. On the other hand, approaches are pursued in which the angle of the interferer is explicitly estimated and a null for interference suppression is then steered in this direction.

13.3.4.1 Measures Based on Angle Estimation Methods

The disturbed radar signal after range and velocity processing is not suitable for determining the interferer's angle because targets are not or hardly detectable due to the interference. For this reason, the interferer's angle is estimated using the disturbed time-domain signal. Since the sought targets are also contained in this, a comparison of the angle estimation of the disturbed with the undisturbed signal components is carried out. The difference gives the angle of the interferer. If the interferer's angle is known, digital beamforming is used to set a null in the disturbed velocity cells in the direction of the interferer before the usual signal evaluation is carried out.

When using MIMO radars, it must be taken into account that no virtual aperture exists for the receive-only mode, i.e. the operating mode in which the interference is received. Consequently, the real receive aperture must be taken into account for the angle estimation of the interference.

13.3.4.2 Adaptive Beamforming

Adaptive beamforming methods for interference cancellation do not explicitly determine the direction of the interferer, but solve an optimization problem using signal statistics. Various approaches are known as optimization goal. For example, the difference between the undisturbed and disturbed signal components can be minimized or the signal-to-noise-plus-interference ratio (SINR) can be maximized directly. The optimizer searches for the optimal beamforming vector by solving the so-called Wiener-Hopf equation [1].

The advantage of adaptive beamforming is that targets are also taken into account when searching for the optimal beamforming vector. However, the computational effort for adaptive methods is higher than for methods that are based on the explicit estimation of the angle of incidence of the interferer.

References

1. J. Bechter, A. Demirlika, P. Hugler, F. Roos, C. Waldschmidt, Blind adaptive beamforming for automotive radar interference suppression, in *International Radar Symposium (IRS)* (IEEE, 2018). https://doi.org/10.23919/irs.2018.8447965
2. J. Bechter, K.D. Biswas, C. Waldschmidt, Estimation and cancellation of interferences in automotive radar signals, in *International Radar Symposium (IRS)* (IEEE, 2017). https://doi.org/10.23919/irs.2017.8008126
3. J. Bechter, F. Roos, M. Rahman, C. Waldschmidt, Automotive radar interference mitigation using a sparse sampling approach, in *European Radar Conference (EURAD)* (IEEE, 2017). https://doi.org/10.23919/eurad.2017.8249154
4. J. Bechter, C. Sippel, C. Waldschmidt, Bats-inspired frequency hopping for mitigation of interference between automotive radars, in *IEEE MTTS International Conference on Microwaves for Intelligent Mobility (ICMIM)* (IEEE, 2016). https://doi.org/10.1109/icmim.2016.7533928

Open Access This chapter is licensed under the terms of the Creative Commons Attribution 4.0 International License (http://creativecommons.org/licenses/by/4.0/), which permits use, sharing, adaptation, distribution and reproduction in any medium or format, as long as you give appropriate credit to the original author(s) and the source, provide a link to the Creative Commons license and indicate if changes were made.

The images or other third party material in this chapter are included in the chapter's Creative Commons license, unless indicated otherwise in a credit line to the material. If material is not included in the chapter's Creative Commons license and your intended use is not permitted by statutory regulation or exceeds the permitted use, you will need to obtain permission directly from the copyright holder.

Index

A
Accuracy, 33, 37, 44–46
  angle, 45
  range, 44
  velocity, 45
Ambiguity function, 155–157
Amplitude Grid Map (AGM), 187–190, 193–195
Analog modulation schemes, 52–69
Angle, 19
Angle estimation, 87, 117, 133–141
  Bartlett beamformer, 137, 141
  Capon beamformer, 137, 141
  comparison, 141
  correlation, 133–134, 141
  Fourier transform, 134–135, 141
  MUSIC, 137–141
Angular spectrum
  Bartlett beamformer, 136–137
  Capon beamformer, 137
  correlation, 133–134
  Fourier transform, 134–135
  MUSIC, 139
Angular velocity, 18
Antenna and Interconnect Technology (AIT), 164–170
Antenna arrays, 41, 117–130, 164
  ambiguity function, 155–157
  array design, 155–157
  beamformer, 125–126
  calibration, 125
  electric field, 119–122
  near-field effects, 157–159
  radiated field, 119–122

receive signal, 123
  covariance, 126
  signal model, 123–125
  sparse arrays, 157
Antenna in Package (AiP), 164
Antenna on Chip (AoC), 164
Antennas, 170–174
  integrated antennas, 172–174
  lenses, 172–174
  patch antennas, 170–171
  waveguide antennas, 171–172
Antenna spacing, 41, 42, 129
Aperture, 41, 118
Atmospheric attenuation, 21
Autocorrelation, 77, 106
Automation technology, 28–29
Automotive radar, 5, 25–27
Autonomous driving, 27, 185
Azimuth, 118

B
Backprojection, 206–208
Bartlett beamformer, 136–137, 141
Beamformer, 125–126, 134–141
Beamforming, 117, 221
  adaptive beamforming, 221
Beat signal, 53
  chirp sequence, 65, 67–68, 95
  CW, 54, 92
  FMCW, 59, 93
Bistatic, 10
Body scanners, 29

C

Calibration, 125, 133
Capon beamformer, 137, 141
CDM-MIMO, 152–153
 unambiguous range, 153
CFAR, 87, 108–114
 Cell Averaging (CA), 111–112
 Ordered Statistics (OS), 113–114
Chebyshev window, 89–91
Chirp sequence modulation, 64–69, 95–99
Clustering, 114
Clutter, 23
Code-division multiplexing, *see* CDM-MIMO
Codes, 78
Code sequence, 78
Constant false alarm rate algorithm, *see* CFAR
Continuous-time signal, 34
Corner reflector, 12
Coupling paths, 180
Covariance matrix, 126, 136, 138
Cramèr-Rao lower bound, 44
Cross-correlation, 77
CW modulation, 52–56, 92–93

D

Data cube, 35
Data fusion, 186
3 dB-beamwidth, 43, 86
Dead reckoning, 196
Delay, *see* Propagation time
DFT, *see* Fourier transform
Dielectric material, 165–166
 conductivity, 165
 loss tangent, 165, 166
 permittivity, 165, 166
Dielectric waveguide, 169
Diffraction, 20, 21
Digital beamforming, *see* Angle estimation
Digital modulation schemes, 69–81
Direct Digital Synthesis (DDS), 175–176
Direction of arrival, *see* Angle
Doppler radar, *see* CW modulation
Doppler shift, 15–19
Duality, 118
Dynamics
 channel, 8
 general, 14
 targets, 14

E

Electric field, 119–122
 distance factor, 122
 element factor, 122
 group factor, 122
Elevation, 118
Embedded Wafer Level Ball grid array (eWLB), 167
Equivalent Isotropic Radiated Power (EIRP), 178
Extended targets, 13

F

Far field, 6
Fast time, *see* Frame
FDM-MIMO, 151–152
 unambiguous range, 151
FFT, *see* Fourier transform
Field of view, 118
FMCW modulation, 56–64, 93–95
Fourier transform, 34
 DFT, 34, 35
 DTFT, 34
 FFT, 35
 IDFT, 71
Frame, 35
 fast time, 35
 slow time, 35
Frequency bands, 4–6
Frequency-division multiplexing, *see* FDM-MIMO
Frequency hopping, 219
Frequency ramp, 58
Frequency regulation, 4–5
FSK radar, *see* Modulation schemes

G

Gaussian distribution, 187
Ghost target, 155
Global Navigation Satellite System (GNSS), 189, 196
Gold sequence, 77
Grating lobe, 122, 155, 157
Grid map, 185–200
 amplitude grid map (AGM), 187–190, 193–195
 free space model, 190–191
 global map, 187
 local map, 187, 190–193
 map creation, 186
 map update, 193
 occupancy grid map (OGM), 190–195

Index

occupancy probability, 190, 191
probabilistic, 190–193
SLAM, 196–200
target model, 191–193

H
Hardware effects, 177–181
  leakage, 178–180
  phase noise, 180–181
Hülsmeyer, 3

I
Integrated antennas, 172–174
Integration gain, 86–87, 178
  chirp sequence, 99
  MIMO, 153–155
  OFDM, 104
  phased array, 153–155
  SAR, 206
Interference, 213–221
  analog radars, 214–216
  chirp sequence, 214–216
  countermeasures, 217–221
  detection, 217
  digital radars, 216–217
  FMCW, 214–216
  OFDM, 216–217
  PMCW, 216–217
Interference impact, 214–217
  analog radars, 214–216
  chirp sequence, 214–216
  digital radars, 216–217
  FMCW, 214–216
  OFDM, 216–217
  PMCW, 216–217
Interference mitigation, 217–221
  adaptive beamforming, 221
  angular domain, 221
  beamforming, 221
  fequency domain, 219
  null steering, 221
  time domain, 218–219
    signal reconstruction, 218–219
    smoothing, 218
    zeroing, 218
  time-frequency domain, 219
Interference signal, 213–214
Isotropic power density, 7

K
Kasami sequence, 77

L
Leakage, 178–180
Lenses, 172–174
Link budget, 177–178

M
Manufacturing tolerances, 166
Mapping, 185
Matched filter, 85
Material parameter, 165
Maximum length sequence, 77
Maximum range, *see* Unambiguity
Measurement dimensions, 14
Medical applications, 31
Micro-Doppler signature, 19
Millimeter wave packages, 167
Millimeter wave range, 4
MIMO radars, 143–159
  array design, 155–157
  integration gain, 153–155
  orthogonal signals, 149–153
  SNR, 153–155
  virtual aperture, 143–148
Modulation schemes, 51–81
  analog, 52–69
  chirp sequence, 64–69
  CW, 52–56
  digital, 69–81
  FMCW, 56–64
  FSK, 56
  OFDM, 70–76
  PMCW, 75–81
  pulse, 51
Monolithic Microwave Integrated Circuit
    (MMIC), 163–164, 167, 174–175
Monopulse, 174
Monostatic, 10
Multipath propagation, 20, 21–23
MUltiple SIgnal Classification (MUSIC),
    137–141
Multiplexing, *see* Orthogonal signals

N
Narrowband condition, 17
Near field, 157
Nyquist-Shannon theorem, 34

O
Occlusion effects, 186
Occupancy Grid Map (OGM), 190–195

OFDM modulation, 70–76, 99–105
  cyclic prefix, 73
  intercarrier interference, 103
  intersymbol interference, 73
  maximum range, 101
  maximum velocity, 103
Orthogonal signals, 149–153
  CDM, 152–153
  FDM, 151–152
  TDM, 149–151

P

Package, 164
Parallax, 11, 174
Patch antennas, 170–171
Path difference, 123
PCB technology, 166–167
Peak detection, 114
Phased arrays, 153–154
Phase-locked loop, see PLL
Phase noise, 176, 180–181
  range correlation, 180–181
Phase noise power density, 176
Phase-Shift Keying (PSK), 71
PLL, 57, 175–176
PMCW modulation, 75–81, 105–109
Point Spread Function (PSF), 43
Printed circuit board, see PCB technology
Propagation time, 14, 53
Pulse modulation, 51

Q

QFN package, 167, 168
Quadrature Phase-Shift Keying (QPSK), 71

R

Radar cross section, see RCS
Radar equation, 6–11
  area targets, 9–10
  point targets, 6–8
Radar hardware, 163–176
  antenna in package (AiP), 164
  antenna on chip (AoC), 164
  antennas, see Antennas
  assembly and interconnect technology, 164–170
  DDS, 175–176
  eWLB, 167
  millimeter wave packages, 167
  MMIC, 163–164, 167, 174–175
  PCB, 166–167

PLL, 175–176
QFN packages, 167, 168
SIW, 168
system on chip (SoC), 164
system partitioning, 163–164
transitions, 168–170
transmission lines, 168–170
Radar network, 27
Radar signal processing, see Signal processing
Radar technology, see Radar hardware
Radial velocity, see Velocity
Radiation pattern, 127, 129
Range, 7, 14–15, 83
  maximum, 8, 9
Range-Doppler coupling, 61
Range-Doppler matrix, see Range-velocity matrix
Range estimation
  chirp sequence, 95–96, 98–99
  FMCW, 93–95
  OFDM, 100–104
  PMCW, 106–107
Range measurement
  chirp sequence, 65–67
  CW, 55–56
  FMCW, 59–64
  OFDM, 75
  PMCW, 78–81
Range-velocity matrix, 83, 85–86
  chirp sequence, 98
  OFDM, 103
  PMCW, 107
RANSAC algorithm, 198
Rayleigh bandwidth, see Rayleigh criterion
Rayleigh criterion, 38, 42, 127, 128
Rayleigh resolution, see Rayleigh criterion
RCS, 7, 12–14
Receive power, 8, 9
Rectangular window, 88–89
Reflection, 14, 21
Relative velocity, see Velocity
Resolution, 33, 35–43
  angle, 41–43, 127–130
    one-dimensional, 127–129
    two-dimensional, 129–130
  range, 38–40
    chirp sequence, 96
    OFDM, 101
    PMCW, 106
  SAR, 205
  velocity, 40–41
    chirp sequence, 97

Index 227

CW, 93
OFDM, 102
PMCW, 107
RF substrates, 166
Robotics, 28, 185

S

Sampled signal, 34
Sampling theorem, 34
SAR, 203–212
   aperture, 203–204
   azimuth dimension, 204
   backprojection, 206–208
   imaging, 205–208
   integration gain, 206
   inverse SAR, 203
   probabilistic, 208–212
   processing, 206–208
   range dimension, 204
   resolution, 205
   SLAM, 212
Scan matching, 199–200
Scattering, 23
Scattering center, 13
Security technology, 29–30
Semiconductor technologies, 174
   CMOS, 174, 175
   GaAs, 174
   SiGe, 174
Sensor coordinate system, 187
Separability, 33, 36–37, 43–44
Side lobe, 86, 88, 89, 155
Side lobe level, 86
Signal processing, 83–114
   angle estimation, 87
   CFAR, 87
   chirp sequence, 95–99
   CW, 92–93
   FMCW, 93–95
   OFDM, 99–105
   PMCW, 105–109
   signal processing chain, 83–84
   target list, 87–88
   windowing, 86, 88–91
Signal processing chain, 83–84
Signal synthesis, 175–176
Signal-to-noise ratio, *see* SNR
Simultaneous Localization And Mapping (SLAM), 189, 196–200
   dead reckoning, 196
   SAR, 212
   scan matching, 199–200

Simultaneous localization and mappling, *see* SLAM
Skin effect, 165
Slowtime, *see* Frame
SNR, 45
   MIMO, 153–155
Sparse arrays, 157
Spatial direction, *see* Angle
Spectrum, 34
Spherical coordinates, 118–119
Steering vector, 123, 124, 145, 159
Substrate-Integrated Waveguide (SIW), 172

Superstrate, 173
Surface roughness, 165
Synthetic aperture radar, *see* SAR
System on Chip (SoC), 164

T

Target detection, *see* CFAR
TDM-MIMO, 149–151
   unambiguous velocity, 150
   velocity-angle coupling, 150–151
Time-division multiplexing, *see* TDM-MIMO
Tolerances, 166

U

ULA, 41, 123, 127, 134, 144–146, 147, 156
Unambiguity, 33, 37, 46–47
   angle, 118, 129–130
   range, 46
     CDM-MIMO, 153
     chirp sequence, 96
     CW, 56
     FDM-MIMO, 151
     OFDM, 101
     PMCW, 107
   velocity, 46
     chirp sequence, 97
     CW, 93
     OFDM, 102
     PMCW, 108
     TDM-MIMO, 150
Uniform linear array, *see* ULA
Uniform rectangular array, 127, 129

V

Velocity, 15–19, 83
Velocity-angle coupling, 150–151
Velocity estimation

chirp sequence, 95–99
CW, 92–93
FMCW, 93–95
OFDM, 102–104
PMCW, 107–108
Velocity measurement
  chirp sequence, 65–67
  CW, 54–55
  FMCW, 61–64
  OFDM, 75
  PMCW, 78–81

Virtual aperture, 143–148
Von Hann window, 89

W
Waveguide antennas, 171–172
Wave propagation, 20
Windowing, 86, 88–91
  chebyshev, 89–91
  rectangular, 88–89
  Von Hann, 89

The manufacturer's authorised representative in the EU is Springer Nature Customer Service Centre GmbH, Europaplatz 3, 69115 Heidelberg, Germany. If you have any concerns regarding our products, please contact ProductSafety@springernature.com

Printed and bound by CPI Group (UK) Ltd, Croydon, CR0 4YY

15/12/2025

02019677-0007